COMPLEX MACHINING AND AI-METHODS

COMPLEX MACHINING AND AI-METHODS

Proceedings of the IFIP TC5/WG5.3 Working Conference on
Process Planning for Complex Machining with AI-Methods
Gaussig, Germany, 27-29 November 1991

Edited by

DETLEF KOCHAN

Technische Universität Dresden
Institut für Fertigungsinformatik
Dresden, Germany

1991

NORTH-HOLLAND
AMSTERDAM • LONDON • NEW YORK • TOKYO

ELSEVIER SCIENCE PUBLISHERS B.V.
Sara Burgerhartstraat 25
P.O. Box 211, 1000 AE Amsterdam, The Netherlands

Distributors for the United States and Canada:
ELSEVIER SCIENCE PUBLISHING COMPANY INC.
655 Avenue of the Americas
New York, N.Y. 10010, U.S.A.

ISBN: 0 444 89326 1

© 1991 IFIP. All rights reserved.

No part of this publication may be reproduced, stored in a retrieval system or transmitted in any form or by any means, electronic, mechanical, photocopying, recording or otherwise, without the prior written permission of the publisher, Elsevier Science Publishers B.V., Permissions Department, P.O. Box 521, 1000 AM Amsterdam, The Netherlands.

Special regulations for readers in the U.S.A. - This publication has been registered with the Copyright Clearance Center Inc. (CCC), Salem, Massachusetts. Information can be obtained from the CCC about conditions under which photocopies of parts of this publication may be made in the U.S.A. All other copyright questions, including photocopying outside of the U.S.A., should be referred to the publisher, Elsevier Science Publishers B.V., unless otherwise specified.

No responsibility is assumed by the publisher or by IFIP for any injury and/or damage to persons or property as a matter of products liability, negligence or otherwise, or from any use or operation of any methods, products, instructions or ideas contained in the material herein.

pp. 85-98: Copyright not transferred.

Printed in The Netherlands

PREFACE

The welfare of Society is dependent on highly productive and high-quality manufacturing systems. These are fulfilling the central need for high productivity in most fields by producing the efficient and highly productive tools needed. Examples such as agricultural machines, computers, electrical generators and transformers can be mentioned.

An increasing demand for customer adaptation and product changes following market shifts are escalating the need for higher productivity in complex machining and for effective, high-quality process plans in complex machining.

The use of computers in manufacturing and planning are fulfilling an increasingly important role in our striving for higher productivity and quality.

In computing, few new technologies have captured our imagination more than Artificial Intelligence. Although significant work in the field was under way more than 20 years ago, only recently has this technology been applied in the CAD/CAM area. Among the candidate areas are those in which human experts perform well, but traditional computer application do not, where the work is too time-consuming and the experts are too few. Highly-productive tools that effectively support and amplify human abilities will then be very important.

Design and manufacturing planning processes for Complex Machining require an expert to develop the technical solutions for a product and for its manufacture. Foremost among the potential benefits of expert systems (an example of artificial intelligence technology) are the capture, improvement, documentation and distribution of human expertise in determining a suitable process sequence, the overlapping of machining operations, the consideration of manifold geometrical, technological and economic rules and new knowledge-based methodology as applied to Complex Machining. Important questions when trying to apply AI techniques are:

- can a scientific and sound base be found within the actual application area?
- can the knowledge be formalized in a computer system?
- can a good human interface be developed?

This working conference has focused on some specific problem areas:

Interfaces between CAD and CAP (based on AI-methods)

- feature definition, recognition of features and feature-generation
- feature definition, generation of operational planes based on product and process information
- machining and process knowledge feedback to design.

Innovations in manufacturing equipment

- Tooling (cutting tools, Tool Dialog systems, new cutting materials and effective selection ...)
- new generation of machining cells and control units
- tool monitoring and machine diagnosis

Modelling and decision support

- decision making for simultaneous engineering
- decision making under uncertainty conditions
- multi-criteria problems
- object and status identification (classification methods)
- Optimization, Branch and Bound.

Intelligent modelling environments

- knowledge acquisition (techniques, strategies for practice)
- knowledge representation (suppositions, means of support)
- shells
- Expert systems in practice.

With limited time and a limited number of papers only a few of these areas have been covered in a reasonably sufficient way. The rest have been left open for future conferences and research work. To those who have contributed this far, we are very grateful.

Torsten Kjellberg
Secretary, IFIP Working Group 5.3

Professor Gus Olling
Chairman, IFIP Working Group 5.3

TABLE OF CONTENTS

Preface   v

Introduction   1

DEVELOPMENT TRENDS IN COMPLEX MACHINING

New Solutions for Process Reliable Automation
H. Hammer (*Invited paper*)   11

INTEGRATED PROCESS PLANNING
Chairman: F.-L. Krause, FhG IPK, Berlin, Germany

The Functionalities: A Continuous Information Flow
between CAD and CAPP
E. Cocquebert, H. Chaouch, D. Deneux and R. Soenen   33

Knowledgebased Process Planning with Object Oriented
Implementation
O. Myklebust   49

An Integrated Software System Supporting a Machining
Cell in Mechanical Engineering
N. Todorov and I. Levi   59

PROCESS PLANNING FOR COMPLEX MACHINING
Chairman: R. Soenen, University of Valenciennes, France

CAPP Based on Advanced Modelling Techniques
H.-K. Tönshoff, M. Becker and J. Kreutzfeldt   67

GENOA: Feature Based Generation and Optimization of
Process Plans
H.-J. Held and G. Jüttner   85

Planning of Operation-Sequences with an AI-Based
Knowledge-Acquisition Tool
H.G. Vogt and P. Zaring   99

OPERATIONAL PLANNING FOR COMPLEX MACHINING
Chairman: M. Szafarcik, Poland

NC-Programming Systems for Multi-Tool Cutting Lathes
A. Storr and Th. Reibetanz — 117

Process-Planning for Complex Machining of Rotational Parts by Means of Knowledge-Based Methods
D. Kochan, M. Hess, J. Oelschlegel and L. Vogel — 129

Feedback for Process Planning as an Approach to Intelligent Quality Assurance
F.-L. Krause, A. Ulbrich and R. Woll — 147

Verification of Knowledge Base in Intelligent Technological Design System
A.F. Kolchin and S.A. Zykova — 161

Increased Productivity and Reliability by Intelligent Software Support
D. Kochan, A. Nestler and Chr. Schöne — 175

MONITORING AND MANUFACTURING ASPECTS
Chairman: A. Storr, University of Stuttgart, Germany

Model Based Diagnostics of Machine Tools and the Turning Process
H. Schulz, H. Schönherr and K.P. Gebauer — 195

A New Integrated Concept for Condition Monitoring and Predictive Maintenance of Machine Tools
R. Sandner — 207

NEW PHYSICAL PRINCIPLES IN MANUFACTURING TECHNOLOGY
Chairman: D. Kochan, Technical University Dresden, Germany

Stereolithography - Fields of Application and Factors Influencing the Accuracy
B.E. Hirsch and H. Müller — 219

PAPERS RECEIVED AFTER DEADLINE

Knowledge Based Process Planning for One-of-a-Kind Production
E. Hämmerle, H. Bochnick, B.E. Hirsch and J. Opas — 237

Problems in Designing Highspeed Milling Tools
H. Schulz, W. Hahner and U. Rondé — 245

INTRODUCTION

1. General development trends

The entire working area of process planning is nowadays characterized by computer support. That means to make it possible to reflect the real technological processes in computer programs it is necessary to build up all manufacturing processes for decision making in detailled models and programs. Because of the whole dynamic technological development such programs and models are more or less of statical kind and therefore not capable to reflect the advanced technological progress in all details. There are some important factors of influence:

- The continuous development of all components in manufacturing processes, such as machine tools, chucks, tools, materials etc.

- the dynamic development in computer-, control- and communicationtechnology as tools for intelligent human activites

- the most important human factor characterized by the feature to collect continuous specific knowledge by intelligent handling and operations. By the scientists of human science this human factor in the socio-technical systems, as they are called, is named "implicite experience knowledge".

The gap between technical modelling implemented in sophisticated computer programs and the real industrial world leading in the last years more and more to such requirements as

- to develop open systems for continuous further extension and modifications,

- to deliver powerful databank systems,

- to allow human intervention in an interactive mode,

It can be pointed out that remarkable progress has been achieved. But independently of this current state of the art it seems necessary to look for new methods and software tools. The AI-methods have a tradition of thirty years development and now first practical results are expected. But no perfect solution for all problem-areas in manufacturing technology can be found. The specific requirements are manifold and no unified solution is possible. For the different working-spheres it is more and more possible to integrate knowledge based methods in the highest form as expertsystems or in another kind of new methodical approaches.

The working conference "Process-planning for complex machining with AI-methods" will contribute to these general developmental trends. Before the content of the working conference is characterized some explaination to the main developmental trend into the direction of "complex machining" may be necessary.

2. Main developmental trends in manufacturing technology

The complex machining has many faces. In this working conference we will emphasize some selected aspects, which mark the current development in computer integrated manufacturing from the technological point of view. Such essential aspects are:

- flexible automation of the entire manufacturing processes for defined classes of workpieces,

- development and application of machining cells for multi-operations,

- further development of manufacturing procedures by improved cutting materials, spindles and controllers, which allow high spead machining,

- new kind of manufacturing technology-called "Solid Freeform Fabrication" or "Rapid Prototyping",

The general trend of complex machining in machining centers, manufacturing cells and flexible manufacturing systems as typical equipment for flexible automation requires a high level of software support. Especially the trend to computer integrated manufacturing requires more and more intelligent and knowledge based softwaresupport.

All these typical and specific directions of complex machining are the background for improved methodical support by different approaches of knowledge based methods.

Process automation

This developmental trend requires the automation of many functions of the entire production process such as:

- machining of different operations,
- tool changing, management and monitoring,
- workpiece handling, storage and transportation,
- supervising, quality control and other auxiliary functions.

The planning for process reliable automation requires a lot of practical experiences. With the invited paper by Mr. Hammer,

Managing Director of The Fritz Werner Werkzeugmaschinen AG an excellent example of the entire process of reliable automation will be presented.

All other included papers carefully selected and evaluated by international referees deal with two main subjects:

- new approaches of methodical aids within knowledge based methods (see point 3.);

- new manufacturing technologies and related new methods for process and operational planning (see point 4.).

3. Knowledge based methods for flexible automation

In accordance with the development of flexible automation there are created powerful software support in kind of NC-programming systems and CAD/CAM-packages for determined problemareas.

The further steps in direction of application of AI-methods leads to first approaches of knowledge based methods. Object of the research work and first results are the entire information flow CAD-CAPP-CAM including monitoring and feedback's, which will be central scope of the working conference.

A team under the leadership of **Prof. Soenen (University Valenciennes, France)** developed a new concept for a dynamic model of the interface between CAD and CAPP. The goal is to fulfil all requirements throughout the essential working phases:

- process planning phase,
- manufacturing phase,
- process control.

Mr. O. Myklebust (SINTEF Trondheim, Norway) offers the theme "Knowledge based process planning with object oriented implementation". The process planning system will cover all basic process planning functions, and makes available a modelling tool for exploiting the open information gateway into the part description model.

Prof. Todorov and I. Levi (CAD/CAM-Center of Machine Tool Institute Sofia; Bulgaria) contributed with an interesting further development of the principles of Group Technology by the paper "An Integrated Software System supporting a machining Cell".

Prof. Tönshoff a.o. (University Hannover, Germany) dealt with on main feature of the CIM-development the feedbacks and correlations between CAPP and CAD. The result is "A new planning strategy based on advanced modelling techniques..."

The cooperation between H.J. Held (Germanys Application Oriented AI Center Ulm) and G. Jüttner (MAHO AG Pfronten, Germany) leads to a global data model, which considered CAD, CAPP-, CAM (NC)-requirements. Artificial intelligence techniques are used for generating process plans for prismatic parts.

The paper of G. Vogt and P. Zaring (Chalmers University of Technology Gothenburg, Sweden) dealt with "Planning of operation sequences with an AI-based knowledge acquisition tool". An essential goal is delivering a decision support system for the design phase dealing with producibility.

Prof. Krause a.o. (IPK Berlin, Germany) dealt with the "Feedback for Process Planning as an Approach to Intelligent Quality Assurance". Based on a deep analysis of different types of necessary feedbacks from the shop floor back-up to process-planning a first step to self learning systems was gone. The primary item is the generatic feedback about process parameters.

A. Kolchin and A. Zykowa (STANKIN Moscow, USSR) deal in their paper with "Verification of knowledge base in intelligent technological design systems". Background are process planning functions for rotational parts. The very deep approach for the application of AI-methods is also a contribution into the direction of self learning operations.

Prof. Hirsch et al.(BIBA Bremen, Germany) presents the paper "Knowledge Based Process Planning for on-of-a-kind Production" that means methodical aids for the highest degree of flexibility.

Prof. Kochan et al. (Dresden University of Technology, Germany) dealt with "Increased productivity and realibility by intelligent software support". Further developments of the long year developed optimization strategy as first steps into the direction of knowledge based methods are demonstrated. In the field of monitoring first application of AI-methods in manufacturing technology was achieved.

New aspects are demonstrated in both papers Prof. Schulz et al. (TH Darmstadt, Germany) "Model based Diagnostics of Machine and Cutting Process" (using system theory principles for modelling the dynamic behavior) and

R. Sandner (University Stuttgart, Germany) "A new integrated concept for condition monitoring and predective maintenance of machine tools" (The generation of the knowledge base is realized by intensive Simulation and a fault dictionary).

4. New manufacturing technolgies and related methods

4.1. Machining cells for multi-operations

New approaches of complex machining were developed in the past years especially in connection with turning cells. The functionalities are extented by:

- multi spindle-processing with different tools,
- multi tool processing for turning operations,
- processing with c and y-axes,
- rear drilling attachment.

The implementation of such complex processes, controlled in 4, 5, 7 or more axes required specific new approaches for process planning. The best utilization of the new installed performance is not to be met by the available NC-programming system.

The contribution of Storr/Reibetanz (multi spindle processing) and Kochan/Oelschlegel a.o. (2x2 axes machining) will present new ways for knowledge based problem-solving.

4.2. More advanced cutting processes

It can be pointed out that high speed machining based on new cutting materials, spindles, controllers changed not only the whole operations, but also the theoretical calculation methods. In connection with the aspect of complex machining it is to be emphasized that a new kind manufacturing processes can be realized. That means instead of the conventional subdivision of in several operations (e.g. turning or milling) finemachining by grinding it is possible to solve such manufacturing tasks in one single operation. This trend is still developing.

A lot of experiences are collected by the Darmstadt University of Technology. Some experiences will be presented by **Schulz, Hahner, Ronde** in their paper "Problems in designing high speed milling tools".

4.3. Rapid Prototyping

The true revolution in modern manufacturing technology is characterized by the direct generation of CAD/CAM-data in complicated workpieces mainly connected with specific application of laser-beams and new materials.

This seems the currently highest level of complex machining. As a general description "Solid Freeform Fabrication" is also used, which means the direct production of threedimensional objects based on 3D-CAD/CAM-data. The first commercialized procedure is named "Stereolithography". It is to be understood that the generalized methods for process planning are under development. But most important are all first practical expierences, which will be presented in the paper by **Hirsch/ Müller** "Factors influencing the accuracy of stereolithography parts".

Prof. Dr.sc.techn. Kochan
Chairman IPC

DEVELOPMENT TRENDS IN COMPLEX MACHINING

H. HAMMER
MANAGING FRITZ WERNER
WERKZEUGMASCHINEN, BERLIN, GERMANY

New Solutions for Process Reliable Automation

H. Hammer

Managing Director, Fritz Werner Werkzeugmaschinen AG,
Unterturkheimer Straße 15-23, D-1000 Berlin 48

1. DEMANDS ON MODERN MACHINING CENTERS AND SYSTEM INTEGRATION

Not only is the desire for shorter process times and smaller lot sizes of importance but also the question of high manufacturing quality and reliability, especially in the case of automated systems. Idle times are coming under more and more criticism within the scope of optimum process structuring, and the demands on performance, speed and precision are growing. The development of cutting materials has also made considerable progress in the last few years, with the result that coated tools, polycrystalline cutting materials or ceramics are being used today. Usually, the efficiency of these tools can be utilized only inadequately by the machining centers (MCs) presently in operation. The speed and power of the driving spindle, in particular, leave a lot to be desired.

A doubling of the presently customary spindle speeds, with barely changed torques, is being called for, which entails a corresponding increase in the installed power of the main drive. The main degree of utilization can be considerably improved with shorter non-productive times. That also entails a substantial in the traverse speed. The higher demands on precision and stability require new design solutions.

The use of modern MCs must, however, also be seen in connection with their integration in flexible manufacturing systems (FMS). The flexibility of such manufacturing facilities is mainly related to in-cycle resetting of both workpieces and tools. Greater productivity is achieved and the operator is freed from the pace of the machine.

This requires a modular system concept with standardized components and tested system control units. Only then is the precondition created for gradual expansion to flexible manufacturing cells (FMCs) and manufacturing systems (FMSs).

Moreover, the rising degree of automation and small operating staff permitted thereby make it necessary to improve manufacturing quality and reliability by taking ever broader measures to optimize the manufacturing process. Not only are improvements in manufacturing reliability and quality through the use of sensors indispensable but also comprehensive attention to technological knowledge at every level of the company as regards the structuring of the process. There is usually no organized updating of the process-design criteria and methods encountered in the operations-scheduling and design departments, nor is there generally any recourse to quality-assurance data or the experience of the machine operator. In this connection it is therefore necessary to have access to information on the process at every level, from the process planning to the operative level, including quality assurance. The precondition for this is optimum integration of the manufacturing system in plant operations from the point of view of information technology.

The new development on an MC and its adapted system components described in the following does not only do full justice to the demand profile but also meets, even now, the expectations foreseeable in the near future. This applies to both the machine components and peripheral system modules.

2. THE NEW CASSETTE SYSTEM CONCEPT

With the innovative cassette system concept, workpieces and tools are supplied by one common transport vehicle. This results in considerable advantages over other system concepts (Fig. 1). The design and type of coupling with the internal cassette magazine of the machine (Fig. 2 and 3) also lead to decisive improvements:
- high transport performance due to a transport speed of 90 m/min and pallet transfer times of 7 sec.,
- machining centers with stationary cassette magazine and continuous-circuit roller conveyor for intermediate storage of tool changing cassettes,
- in-cycle, fully automatic exchange of tools by way of tool changing cassettes in the event of job changes and wear,
- computer-based tool management with integrated tool presetting device and tool store as well as loading station for the changing cassettes,
- integrated cell control system with shop-friendly, colored-graphics control station for computer aided system guidance.

In the case of the tool supply system, it is necessary to distinguish between the internal machine tool change specified by the sequence of operations and the external tool change in the event of job changes and wear. The internal

SYSTEM CONCEPT			EVALUATION	
Layout of the manufacturing system	Workpiece inter-linkage	Tool inter-linkage	Cost percentage system-peripherals	Automatic job change
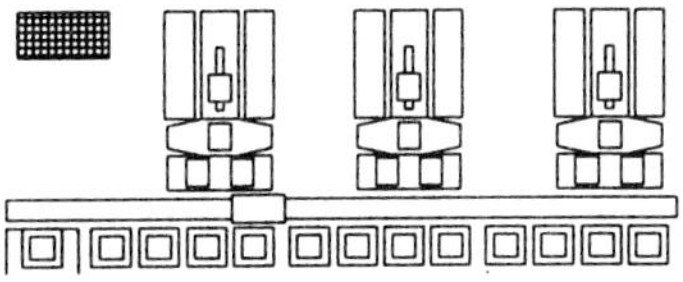	Rail-mounted transport system	None	●	○
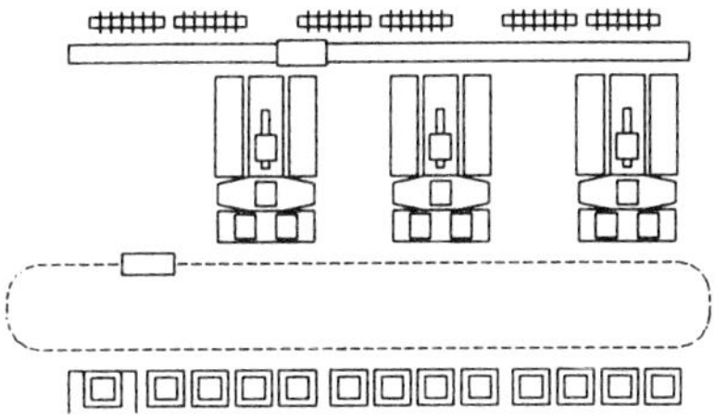	Driverless transport system and storage system	Rail-mounted manipulator	◐	●
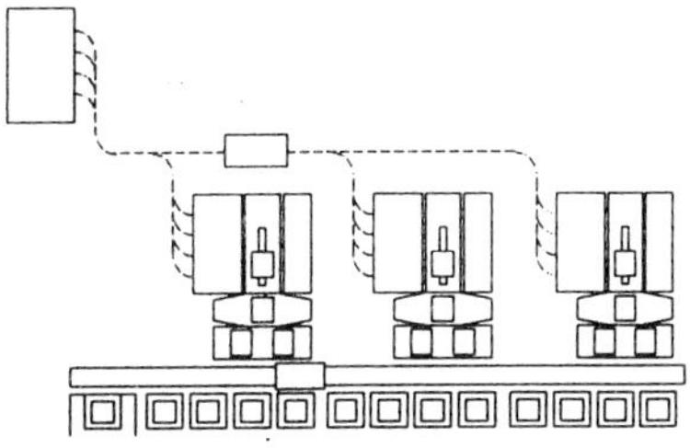	Rail-mounted transport system	Driverless transport system	○	●
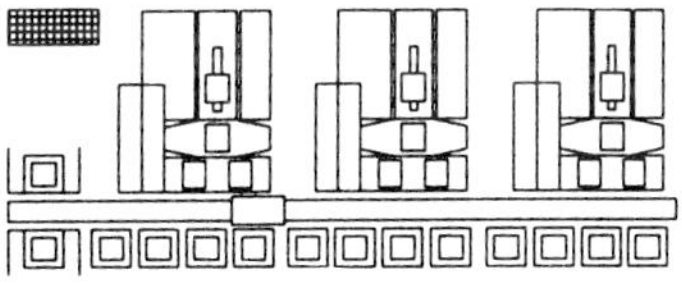	Common rail-mounted transport system		●	●

● good   ◐ average   ○ poor

Fig. 1   Comparison of different system concepts

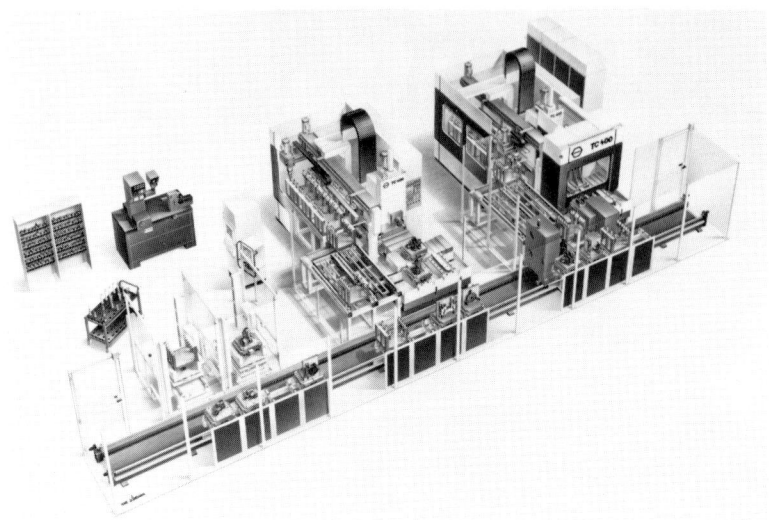

Fig. 2   FMS 400-2 Flexible Manufacturing Cell with common transport system for the supply of workpieses and tools

tool change between the work spindle and cassette magazine takes place within 5 sec. via a tool changer with double gripper and a linear gantry decoupled from the changer. For the external exchange of tools in the event of job changes it is possible for as many as four tool changing cassettes, each with up to 12 differentiated tools, to be placed in intermediate storage on the machine's continuous-circuit roller conveyor. The exchange of tools between the cassett magazine and changing cassettes take place parallel to the machining time by way of the machine's linear gantry. In the front section of the continuous-circuit roller conveyor, the so-called changing area, there is a maximum of two changing cassettes respectively, directly accessible to the linear gantry (Fig. 3).

All in all, up to four changing cassettes can be respectively stored on the input and output belt. As a result, the changing operation proper at the machine is chronologically decoupled from the transport system and the system operator.

Basically, the cassette magazine holds only the tools required for machining of the current job mix. As soon as the data on an impending job change are known, they are entered in the cell control unit by the operator via an input form on the screen. The tools additionally required

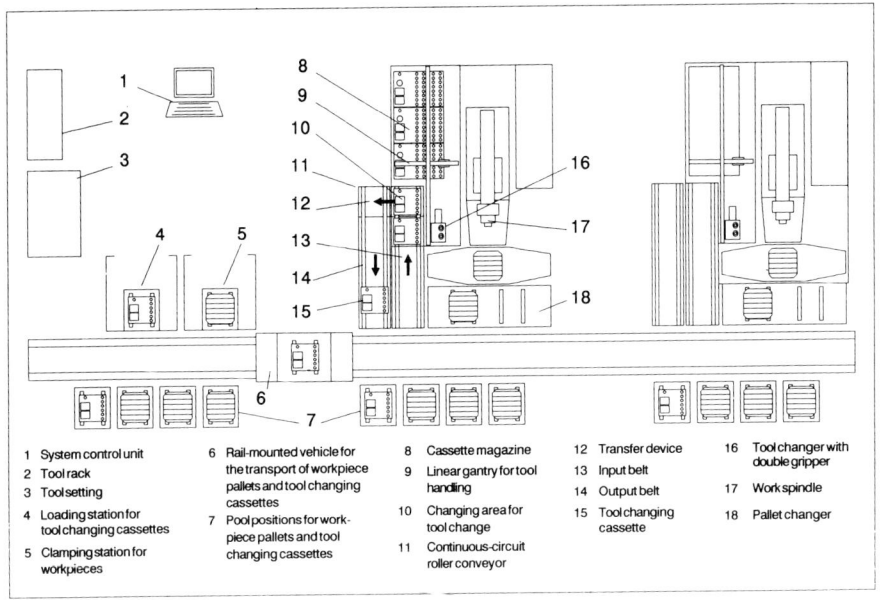

Fig. 3   Layout of the flexible manufacturing cell FMS 400-2

for the following machining and those no longer needed are determined by the computer control system, with attention given to the remaining tool life in the form of a differentiated tool list. All tools to be removed are marked with removal status in the CNC tool file, and a check is made at the same time to determine whether there is space for the new tools to be placed in the cassette magazine. If this is not the case, the operator must provide for another job mix and repeat the input. To avoid additional sorting operations, tools no longer required in the event of a job change are deposited in the changing cassette immediately after use in the spindle, and newly required tools are taken directly from the changing cassette and placed in the work spindle. From the current tool and magazine position codings the control system knows which tools to be removed are still in the storage area and have to be moved to the changing cassettes, and also how many changing cassettes are required for the removal and issuing.

Numerous studies have confirmed that in this way the tools can, as a rule, be changed without downtimes. Only in the case of large numbers of tools and in special cases are additional sorting runs necessary.

In the tool preparation area next to the changing-cassette loading station there ist also the tool store with the preset tools and tool presetting device. The readying of the tools for all the machines integrated in the system is carried out by the operator under computer guidance. The loading and unloading list contains both the storage locations and destinations. This computerbased tool management leads to considerable simplifications and advantages:
- computer aided determination of current tool needs and timely output of preparation lists, with attention given to the remaining tool life,
- faster automatic exchange of tools in the event of job changes and wear due to automatic transport and timely readying of the replacement tools,
- computer aided management and updating of tool data and direct transmission to the machines' CNC control system,
- timely, computer-based reconditioning of tools and intermediate storage through the integration of a presetting device and readying store.

The transport system must not only transport the tool changing cassettes but also supply the workpieces. The pallets loaded with unmachined parts at the clamping and set-up station are automatically transported to the machines and returned to the clamping station after machining. Integrated pallet-storage positions free the operator and transport vehicle from the working pace of the machines and permit storage of a work reserve for runs during breaks and shifts manned with a small staff.

The storage and monitoring of the flexible manufacturing cell is seen to be the modular SC I cell control unit based on an industrial PC. Thus, not only is tool management provided but also all the other advantages of computer-based system operation, like job management and planning as well as production data acquisition and evaluation.

3. SYSTEM COMPONENTS

3.1 Machining Center

In addition to the system peripherals described, the newly developed, compact TC 400 machining center, above all, is highly remarkable/1/. It is a high-performance, horizontal machining center with a pallet size of 400 x 500 mm (Fig. 4). The machining center is prepared for integration in a system and is therefore provided with standardized interfaces on the workpiece, tool and control side. The main features are:

- high efficiency,
- short non-productive times,
- large, expandable tool cassette magazine,
- in-cycle, external exchange of tools, with no setting-up time, via changing cassettes in intermediate storage on the machine and
- high precision.

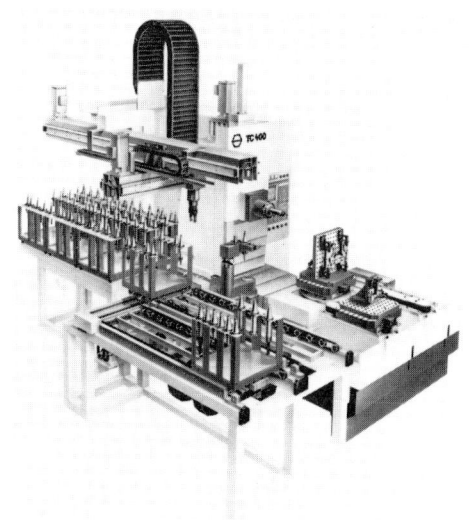

Fig. 4   Compact, system-compatible, TC 400 machining center with a pallet size of 400 x 500 mm

The development of new cutting materials and the machining of aluminium parts require from modern machining centers a high spindle speed with barely changed torques. The drive is thus provided by a newly developed, watercooled, hollow-shaft AC motor with a drive power of 30 KW at 60 % c.d.f. and a torque of 300 Nm at speeds as low as 1000 rpm as well as a maximum speed of 12.000 rpm. This means that an optimum range of power is available for all sorts of machining tasks. Moreover, the rigid design of the spindle drive unit, the thermally symmetric construction and a novel, compact guide system also permit maximum constant precison.

The tool magazine consists of two fixed tool cassettes with 80 positions together, loosely mounted on a base frame. It can be expanded by as many as three cassettes and thus to 200 positions (Fig. 4).

The internal tool change between the magazine and work spindle takes place by way of an NC-controlled, threeaxis linear gantry and a fast changer with double gripper arm that is decoupled from the manipulator. In the transfer position both grippers of the tool changer are within the access range of the manipulator.

The next tool required for machining is deposited by the manipulator in the free position of the tool changer's double gripper; the tool in the other gripper is placed in the cassette magazine right after that. Thus, the tools can be removed from and placed in the changer without the manipulator being tied to the tool changing cycle of the work spindle. This is done at very high travel speeds of the manipulator (90 m/min), so that the shortest tool changing times and chip-to-chip times (8 sec.) are possible.

For the external exchange of tools there are changing cassettes that are likewise loaded and unloaded by the manipulator parallel to machining time. In the event of job changes and wear the tools can be changed either manually or, in the case of interlinked machines, automatically by cassette changes as described.

Short non-productive times are achieved not only with fast tool changes but also with short pallet changing times. Pallets are changed within 12 sec. by a leftright changer accomodated in front of the machine. The pallet changer is also the interface with the transport system.

3.2 Workpiece and Tool Supply System

The special feature of the new system concept is the use of the transport system to supply both workpieces and the tools.

For this purpose, use is made of a numerically controlled, rail-mounted transport vehicle that is distinguished by high functional reliability, short acceleration times, high travel speeds and short positioning and transfer times.

The machining centers are connected to the transport system on the tool side via the continuous-circuit roller conveyor and on the workpiece side via the pallet changer. As many as four tool changing cassettes can be respectively placed in intermediate storage on the input and output belt.

The workpiece carriers are buffered at storage positions along the transport system. If necessary, tool changing cassettes can also be held in intermediate storage at these storage positions.

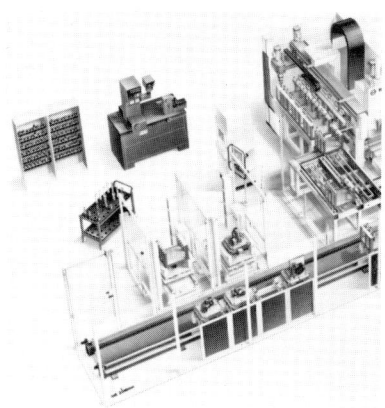

Fig. 5   Tool reading area with cassette loading station, tool store and tool presetting device

The tool area, with integrated tool store and tool presetting device, is linked to the transport system via the connected setup stations (Fig. 5). The tools are transported with 12 changing cassettes that can respectively accomodate up to 12 normal-size and oversize tools. For common transport the changing cassettes have the same dimensions and interfaces as the workpieces pallets.

Use of the transport system for a supply of both workpieces and tools usually does not lead to any bottlenecks, since the transport performance is very high at a travel speed of 90 m/min and transfer times of 7 sec. Moreover, the additional load due to transport of the tool changing cassettes is relatively small, even in the case of frequent tool changes (Fig. 6). Numerous simulation studies have shown that no bottleneck is encountered even in the case of four or more machines and short pallet processing times as well as frequent job changes (Fig. 7)

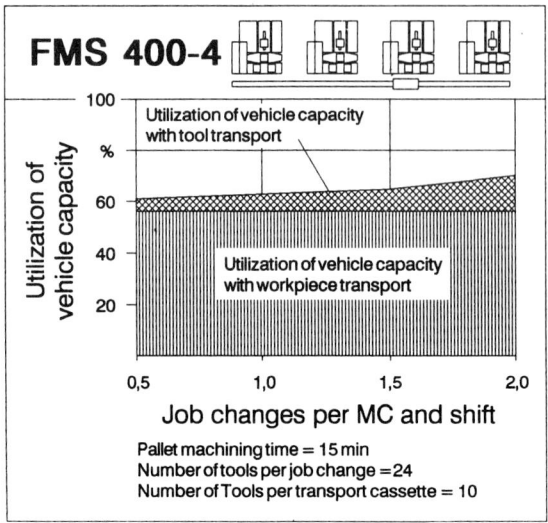

Fig. 6   Utilization of vehicle capacity with workpiece and tool supply system

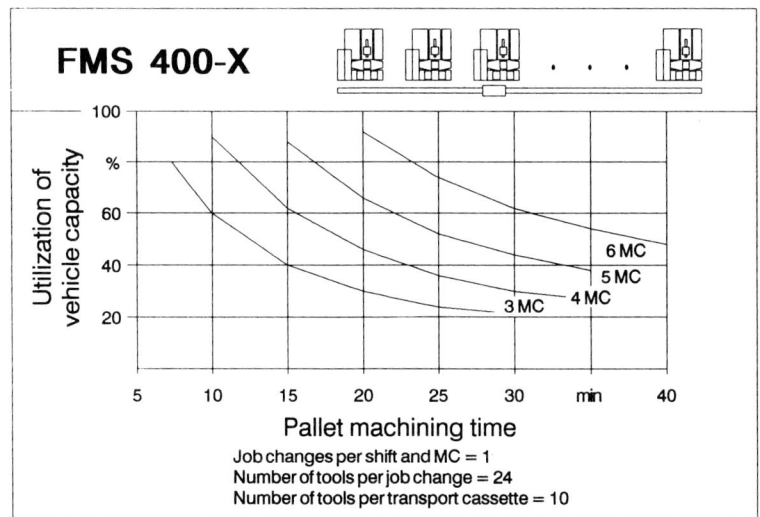

Fig. 7   Utilization of vehicle capacity with different number of machining centers and different pallet process times

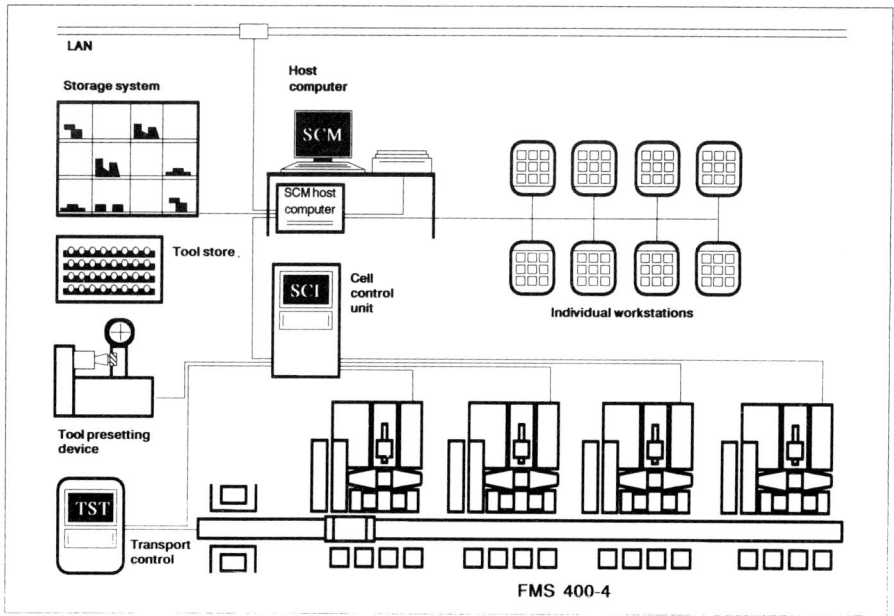

*Fig. 8   Control configuration for the FMS 400-4
flexible manufacturing cell*

3.3   System Control Unit

3.3.1 Hierarchic Control of Interconnected Computers

The functions of the manufacturing cell are controlled and monitored by autonomous control components in a system of interconnected computers (Fig. 8)/2/. The SC I cell control unit based on an industrial PC coordinates and monitors the components of the manufacturing cell, manages the current job and working-stock data and generates control commands that are executed and automatically monitored by the lower-ranking cell components like the CNC control system of the machining centers or the transport control system. Moreover, the process data have to be managed, and alarms from the process monitoring system have to be received and displayed.

If, in addition, there are other manufacturing cells, individual machining stations or storage systems to be controlled at the same time, and if the entire job scheduling, including working-stock utilization planning, is to be optimized, this should be done at a higher-ranking computer level by the SCM host computer. It optimizes the entire job process, with attention given to the availability of workpieces, fixtures, tools and test equipment, initiates readying orders for working stock and supplies information on the current flow of operations and utilization of system capacity. The host computer receives the respective jobs for a limited planning horizon directly from the higher-ranking production planning and control system (PPC).

What is distinctive of this open, modular and hierarchically structured system of interlinked computers is the fact that respective lower-ranking levels can always perform control functions on their own responsibility as long as the framework data transmitted from the higher level are executed and/or complied with. The clear assignment of tasks to autonomous functional groups permits gradual expansion of the system, from a single machine to a complex interconnected manufacturing system.

3.3.2 Cell Control Unit

The basic functions of the SC I cell control unit are job management and planning, NC program management and distribution, production data acquisition and evaluation as well as, above all, the entire tool management. For this purpose the cell control unit also manages all the data on tools in the machine-bound cassette magazine and also the connected readying store as well as the tools at the readying station. The current occupation of the tool magazines can be called up at any time. Bar charts show the remaining tool life still available. In addition, there is a display of how many workpieces can be produced with the remaining tool life available. The current state of the tools can be visually displayed bay way of graphics. The tools needed for the manufacturing jobs planned are calculated with due attention being given to the tools on the machine and the remaining tool life.

An important function of the cell control units is, above all, timely preparation of the differentiated tool list required for the tool change, with indication of the tools additionally required or tools to be removed. The latter are reported directly to the CNC. At the same time, a loading list with the tool number, removal position of the tool in the store and the number of the position in which the tool is to be deposited is printed out for the selective readying of additionally required tools.

For the unloading of tools, an unloading list is printed out that contains the tool number, removal position and number of the storage position in which the tool is to be **deposited again. Indication of both the removal location and** deposit location ensures that work can be done dependably even without readable tool coding. In addition, for reliable identification of the tools outside the machine it is also possible to use adhesive bar-code labels with the place of removal and destination in plain text, data carriers with the code number or programmable data carriers with all the tool data.

The tool presetting device is directly connected to the cell control unit for the transmison of tool data. Measured setting values can thus be directly sent to the tool file and automatically transmitted to the CNC in the case of a tool change. As a rule, the tool store will also be directly included in the flow of information.

Constant acquisition of all production data is likewise one of the basic functions of the cell control unit. Thus, the operating data acquired can be directly evaluated and graphically displayed. Current queries about technical and organizational downtimes, utilization times, production and job-machining times as well as the causes of failures are therefore constantly available.

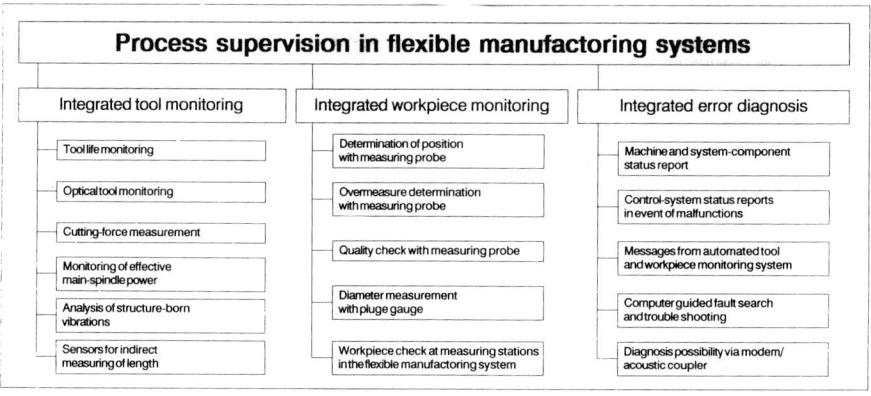

Fig. 9   Process monitoring tasks in flexible manufacturing systems

For the automated supply of workpieces and tools the cell control unit additionally manages the pallet data and travel orders depending on the specified, job-related sequence of operations. For this purpose the sequence of operations is entered via the screen input forms before the job is started or transferred directly from the host computer. The travel orders for the changing cassettes are likewise generated and their execution monitored on the basis of current tool needs.

3.4 Process Monitoring

The process reliable operation of a flexible manufacturing system with, in part, low-supervision production times necessitates comprehensive process monitoring consisting of (Fig. 9):
- monitoring of the tools used for drilling and milling,
- measuring of workpieces in the machine's working space prior to machining,
- quality control of the workpieces produced,
- function check and error diagnosis of the individual system components.

4. INTEGRATION IN THE PLANT ENVIRONMENT

4.1 Replacement of Worn Tools

In addition to the normal exchange of tools in the event of job changes, it must also be possible to automatically replace worn tools. All tools with lives below a prewarning limit are deposited in the changing cassettes and the same displayed to the operator on the control panel. If no sister tools are available, a warning is issued in good time.

For the replacement of worn tools a changing cassette loaded with the reconditioned tools is transported to the machine and exchanged for the changing cassette with the worn tools. Broken tools can additionally be exchanged by hand via a drawer.

Due to setting of a prewarning limit shortly before the end of tool life it is possible to achieve much greater planning latitude for the readying and exchange of worn tools. As a result, it is no longer necessary to provide for a large number of sister tools in the machinebound cassette magazine. Instead, the sister tools can now be placed in the magazine in selective and timely fashion, which considerably reduces tool inventory.

4.2 Tool Identification

The precondition for good tool management is not only a functionally designed tool logistics system but also assurance of optimum tool data flow. The tool data, consisting of the tool number, geometric and technological data as well as the designation of the type of tool, are directly transferred inside the manufacturing cell between the tool presetting device and the cell control unit as well as the CNC system of the machining center.

For certain identification and better tracking of tools outside the tool cassettes and in the tool store it is possible to use not only the computer-produced loading and unloading lists but also, if necessary, adhesive bar-code labels with the tool's place of removal and destination in plain text. Moreover, certain identification of the tools can also be provided by data carriers with fixed coding or programmable memories. A disadvantage in this connection is the fact that every point involved in the exchange of data must be equipped with a read/write station of its own, and every tool requires a data carrier. The costs entailed by the realization of this solution must therefore be seen in relation to the higher data security.

5. INTELLIGENT PROCESS PLANNING

With the availability of powerful CAD/CAM systems the use of computer-aided planning methods is becoming more and more important in the design and job-scheduling departments upstream from the production department. CAD-supported descriptions of even very complex component geometries belongs just as much to the state of the art today as the direct preparation of the NC program on the basis of CAD data. But the treatment of the data relevant to the manufacturing technology is largely left to the programmer's experience in the fields of design and operations scheduling. And the integration of technologically oriented system modules is gaining acceptance only very slowly in industrial applications. Another important aspect is safeguarding the individual employee's know-how and the increase in the availability of process knowledge within the plant entailed thereby.

Basically, there are the following different approaches to computer-aided process planning:
- technology data bases,
- rules expressible in the form of algorithms,
- process models and
- expert systems.

The simplest method, and thus the most wide-spread one, is the use of data bases for the management of proven process variables, examples of successfully handled production tasks as well as process- and plant-specific guidelines for the design of the process. A precondition for effective utilization of such empirical data bases is not only their complete integration in the CAD/CAM environment but also a user interface that is matched to the respective manufacturing process and adequately supports the necessary and regular updating of the stored know-how.

Knowledge of process-technology relationships is distinguished in many cases by rules and guidelines specific to the process or company. When applied to a practical case, they follow a fixed structure of events and produce targets for the process design. The spectrum of approaches already realized ranges from technologically oriented manipulation of geometries in the CAD system to complex modules that start with a description of the processing task (workpiece geometry, material, quality requirements) and provide suitable process variables and process-guidance strategies.

Another kind of technology module is based on process models that make it possible to describe the process structure. Depending on the complexity of the models used, it is possible to ascertain suitable variables for the production tasks at hand or to simulate the course of the process in order to check the given process parameters.

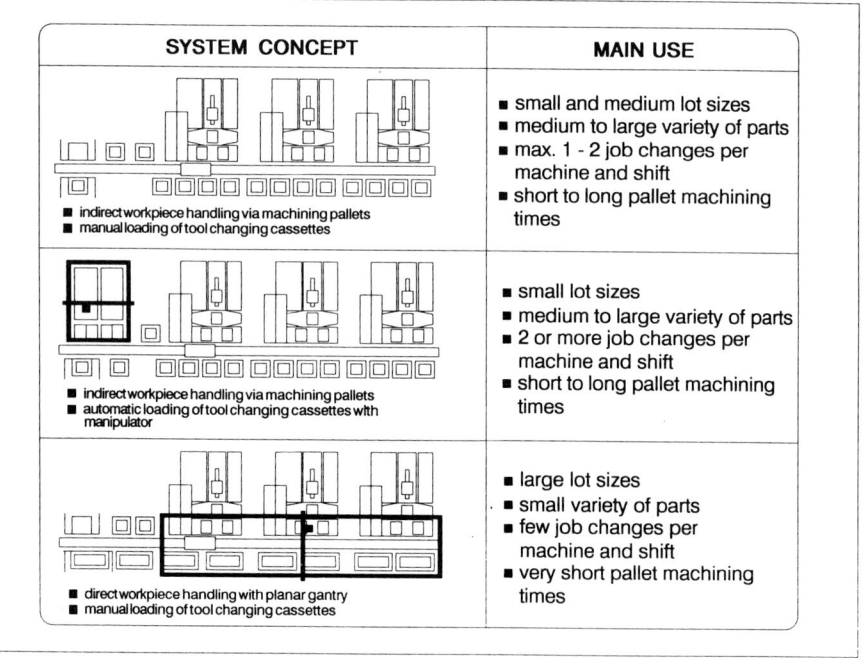

Fig. 10 System variants for special applications based on a modular system concept

The highest state of development in regard to modules for a computer-aided design of processes is represented by expert systems, which can be viewed for simplicity's sakes as a linking of data bases, sets of rules and process models. However, there are hardly any examples of industrial use for the design of processes. The expert systems developed so far are must be viewed exclusively as prototypes. The great expense entailed in practice by the introduction of an expert system - mainly as regards the integration and system-oriented handling of process knowledge - must be viewed as one of the main reasons for the fact that they have been put to so little use hitherto.

6. EXPANSION STAGES

Other system variants tailored to the respective special case can also be realized on the basis of the system concept presented (Fig. 10). Thus, automated loading of the tool changing cassettes via a planar gantry is possible, for example, in the case of extremely frequent job changes per machine and shift. On the other hand, the workpieces caon be directly supplied via a planar gantry on Euro-pallets when there is a small variety of parts, large lot sizes and short pallet processing times. Determination of the most expedient respective system configuration requires careful holistic planning of all alternative solutions. The system should subsequently be tested and optimized with the help of a simulation study. In this way it is possible to select and specify the most functionally suitable and economical configuration for the respective application.

7. BENEFITS AND ECONOMY

Compared with individual machines, important application benefits are conferred by the flexible manufacturing cell described, due to the automated supply of workpieces and tools as well as the computer control:
- greater utilization of machine capacity,
- greater flexibility,
- production with JIT strategy,
- greater transparency.

These improvements can lead to considerable cost cuts. Moreover, great reductions in piece times are achieved due to the greater machine power and higher speeds of the newly developed machining center. The cuts in manufacturing costs possible with the manufacturing cell amount to some 15 to 25 %, depending on the application and size of the system

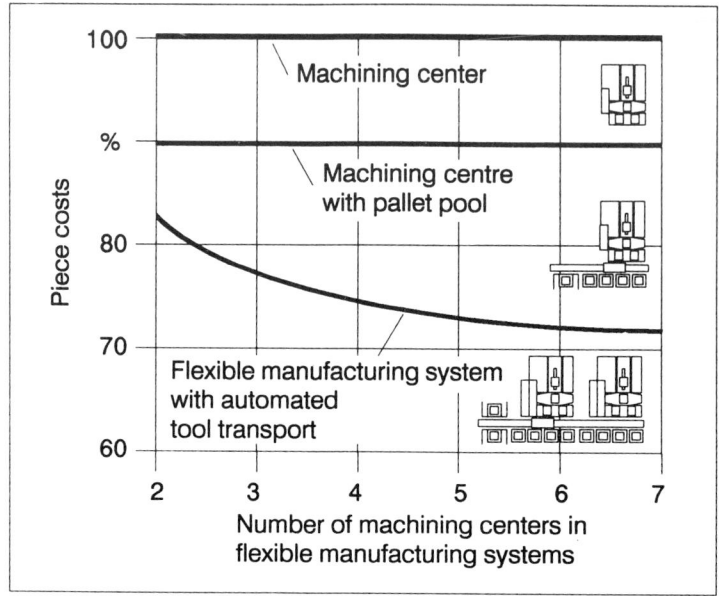

Fig. 11   Economy of the flexible manufacturing cell compared with alternative system concepts

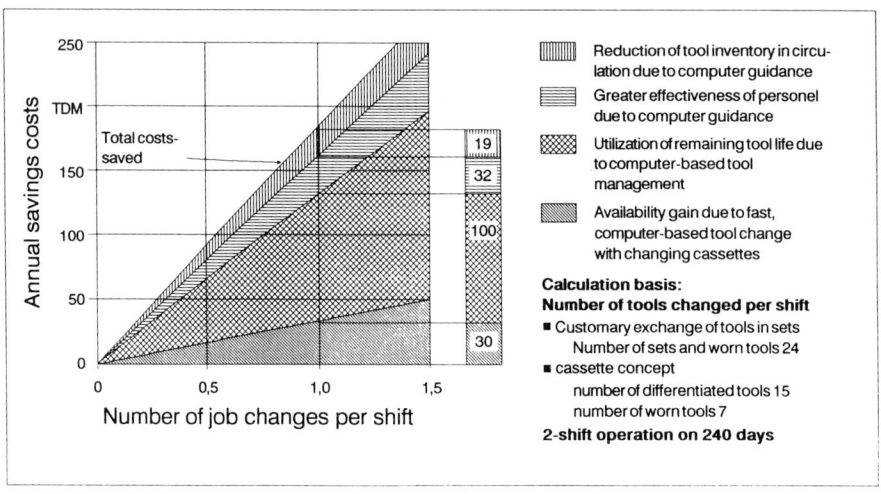

Fig. 12   Potential cost cuts through computer-based tool management

(Fig. 11). The cost-cutting effect rises, above all, with the number of integrated machines, since the costs for the system peripherals drop proportionately.

But costs are cut mostly by computer-based tool management. In comparison with the manual exchange of tools hitherto customary, an automated, in-cycle supply of tools leads to a roughly 10 % improvement in machine utilization with an average of one job change per machine and shift. In addition to the higher productivity achieved with the gain in availability, complete utilization of remaining tool life also leads to considerable cost cuts (Fig. 12).

Due to the computer control, tools are only reconditioned when the entire tool life has been used up. Furthermore, the automated supply and planning of tools requires much less labor and leads not only to a reduction in the tool inventory in circulation but also to much fewer organizational disruptions. Fig. 12 elucidates the annual, potential cost cuts that be achieved with computer-based tool management.

In summary it must be noted that the novel system concept presented can now be used to meet in full the high technical and economic requirements of FMS applications in the case of small machining centers as well. Thus, it is now possible to take advantage of flexible manufacturing in this area as well.

LITERATURE

1. H. Hammer: Neue Lösungen zur flexiblen Automatisierung von Bearbeitungszentren. Werkstatt und Betrieb, 1989, No. 8

2. H. Hammer und B. Viehweger: Einsatz von Leitrechnersystemen für flexibel automatisierte Produktionsanlagen. wt-Werkstattstechnik 79 (1989), No. 8

3. Flexible Fertigungssysteme in der Praxis. Planung, Einführung, Betrieb, Wirtschaftlichkeit. Werner und Kolb, Berlin 1988.

Illustrations: FRITZ WERNER Werkzeugmaschinen AG, Berlin

INTEGRATED PROCESS PLANNING

CHAIRMAN: F.-L. KRAUSE
FhG IPK, BERLIN, GERMANY

THE FUNCTIONALITIES : A CONTINUOUS INFORMATION FLOW BETWEEN CAD AND CAPP

E. Cocquebert, H. Chaouch, D. Deneux, R. Soenen

Université de Valenciennes et du Hainaut-Cambrésis
Laboratoire de Génie Industriel et Logiciel
(URIAH - UA CNRS n° 1118)
Le Mont-Houy
59326 Valenciennes cedex - FRANCE

Abstract

Many works all over the world are led to achieve integration of the product life cycle. Concerning the CAD step, the research aims to define one model which would take into account the minimal set of features of a product. Concerning the CAPP step, works are either closely coupled with the research concerning the CAD step or not. Nevertheless, these works do not question the philosophy of the CAD systems, instead, another research axis (quite recent) aims to capture the design intent. This paper, concerning this research axis, proposes the concept of functionalities as a solution to capture the design intent, taking into account the machining of the designed shapes.

1. INTRODUCTION

Many works all over the world are led to achieve integration of the product life cycle.

Concerning the CAD step, different typologies appeared to help classify the characteristics of a product. Marks proposed five categories including the minimal set of information absolutely necessary for product analysis [1]. They are named geometric features, precision features, technological features and administrative features. From this classification, various kinds of models were defined. We distinguish here the integrated model and the homogeneous model.

The integrated model is a framework made of different models in charge with respectively the different types of features which are form features, precision features and manufacturing features. This model uses both CSG and B-Rep modelizations [2] [3] [4] [5] [6].

The homogeneous model aims to unify the different types of features of a product (geometry, dimensions, tolerances) under a single representation. This model uses geometric reasoning [7] [8], high and low level features [9] [10] [11] [12] [13] [14] for a broader use of the system.

Concerning the CAPP step, the research work is highly dependant on the way the features are used during the CAD step.

The actual philosophy of CAD/CAM systems may be kept, i.e. some tools to design with primitives, boolean operations, geometric transformations, form features,... are provided to the user. This requires a compulsory post-processed translation (recognition and extraction)) to match specific application requirements (eg : process planning) [2] [3] [4] [12] [13] [15].

Another solution is to design exclusively by features in order to ease the feature extraction process using destructive solid geometry. This approach totally follows the design for manufacturability philosophy (the features are process planning oriented) [6] [9] [16].

The previously exposed results may be perceived as works which define a product model from CAD step towards CAM step through CAPP step. Research teams have been working on CAD by features and are still presently doing so.

We think it is interesting to mention some other results concerning more specifically the automatic process planning using AI techniques to generate process plans. We can mention GARI [17] which plans the process for milled parts and which has been followed by PROPEL [18] and GAGMAT [19] ; TOM [20] which concentrates only on hole-making processes and which considers one hole at a time and so on [21].

It is interesting to notice all these works do not question the philosophy of CAD/CAM systems : provide the user with a number of efficient tools to design parts (CAD) and develop treatments of information issued from the CAD step (CAPP). That is why it is now important to talk about a recent research axis : how to capture the design intent ?

This research axis deals with the relationships between functionality and features : the reason why an entity exists, as well as what the requirements of the entity are [22]. This work is required because :
- the taking into account of the designer intention allows to explicit the bright volumes from functional parameters,
- since these volumes are those which effectively have to be machined, there are less translations. Moreover, it is easier to determine the global process plan from the knowledge of the machining process of these volumes.

Thus, from a general point of view, it is presently more important to study integration of the design intent in one product model at the CAD step, in order to obtain a better communication with CAPP step (and obviously the CAM step) [23] [24] [25] [26], or for other downstream applications [28].

This paper is concerned by the following question : "How to integrate the design intent in the product model" and proposes the concept of functionalities as a solution. Firstly, the context of the research in Valenciennes is presented : the functional system of product working out is described as well as the hearth of this system : the dynamical product working out model. Before detailing the use of the concept of funtionalities, we talk about its bases : the design philosophy and the taking into account of the machining process. An originality of this concept is that the user reasons firstly on the functional faces of the product before the shape of the parts within the product. An example illustrates in the last part how this concept is used. In conclusion, we recall the expected advantages of this concept, and the researches we have in view.

2. THE PRODUCT WORKING OUT MODEL

2.1 Functional system of product working out

The homogeneous modelling CAD/CAM system we are planning to develop must fulfil the requirements of the different phases of product working out. It is consequently subject to the successive sollicitations of the overall and definite conception phase, the process planning phase, the manufacturing phase and lastly the process control. Therefore, the product model is made of a dynamical model, since it permanently evolves, through a stepped valuation of both geometric and technologic descriptions. As it is shown in figure 1, the dynamical model is under the influence of:
- The functional model : it is able to translate to the system the designer's intent and the abstract view he has about the product at this early step of the process. The designer expresses his intention in terms of functions to be met, external sollicitations,...
- The object model : it contains some generic representations of various entities (geometrical entities, dimensions and tolerances entities,...). There is a close dependancy of both functional and object models since a given function may induce a certain geometry.
- The production model : it links the technological entities with their corresponding machining cycles. There is also a close relationship of both production and functional models since a given geometry may induce machining parameters, machining tools, NC tool path,...

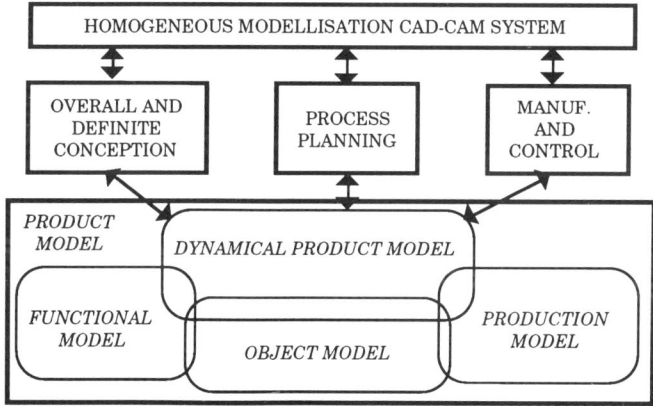

figure 1 : functional system of product working out

We shall now detail the cyclic process which takes place within the dynamical model we have mentionned.

2.2 Dynamical product working out model (DPWM)

Drawn from Kimura's proposal [25], it takes into account the different required steps to work out a product model and represents the one and only interface between the user and the system (figure 2).

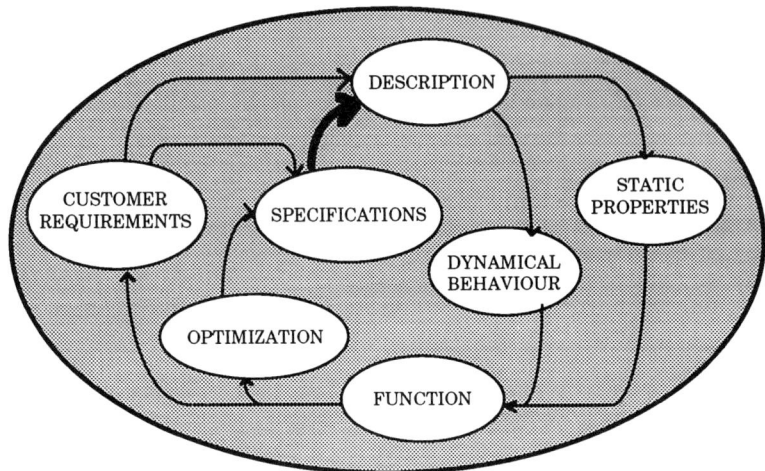

figure 2 : dynamical product working out model

At the origin are traditionally the contractual conditions to be found. They represent the customer's requirements and prescribe the designer a certain list of constraints he has to translate into an adequate geometry garanteing the lowest production cost. This geometry will lead to a description, which at the first motion will be succint. The required functions are also to be translated into specifications, such as a functional scheme or a given choice of features that will possibly modify the original description.

Some experiments on the draft description of the product model can already be carried out to evaluate its static, kinematic and dynamical properties. This helps measure the performance range of the expected product with regards to the specified functionalities.

At this step, a number of methods is available to identify the design mistakes and correct them through modifications of the specifications or by reconsidering the original contractual conditions with the customer.

Having briefly exposed the context of our research (functional system) and the hearth of this model (dynamical product model), we can now present the concept of funtionalities. It is based on :
- the design philosophy which aims to capture the design intent (both functional and object models are involved),
- the taking into account of machining of shapes (both the production and object models are involved).

3. THE CONCEPT OF FUNCTIONALITIES

3.1 The design philosophy
As it is shown in figure 3, the design philosophy is cyclic.

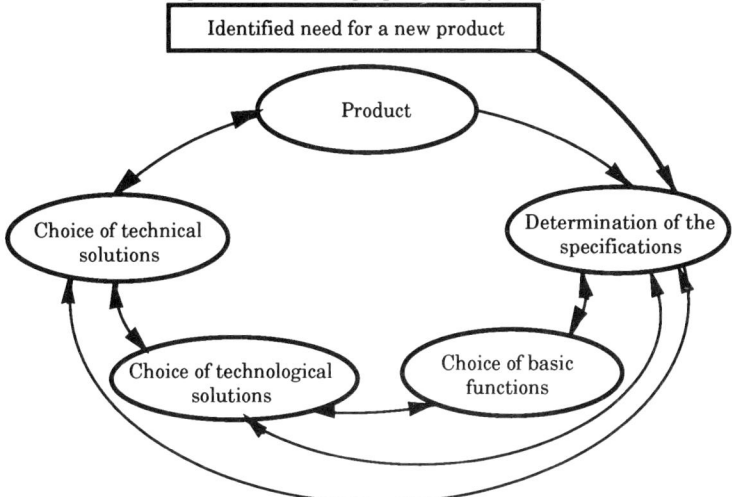

figure 3 : cyclic design philosophy of one product

3.1.1 Identification of the specifications
From an identified need for a new product, the first step consists in detailing its specifications in order to state its general conditions of use. This step is very important because the design of the product is based on these specifications and its validation is also performed according to these specifications.

3.1.2 Choice of basic functions
After detailing the specifications, the design office translates them in terms of basic functions. The designer identifies the different elementary relations (or joinings) between the parts of the product. We can assert that this identification of basic functions is a translation, since the specifications may express an unsolvable need with a current language, conversely to the basic functions which show the feasibility of the design thanks to a specialized language.

In the following, a *basic function* is a joining in which the formulation indicates the authorized or prohibited motions without any precision about the means.

The most often defined basic functions are : complete joining, guiding in translation, guiding in rotation, helicoïdal guiding, axis joining. As it is shown in figure 4, a standard symbol (based on NF E04-016) may be associated to each of these functions.

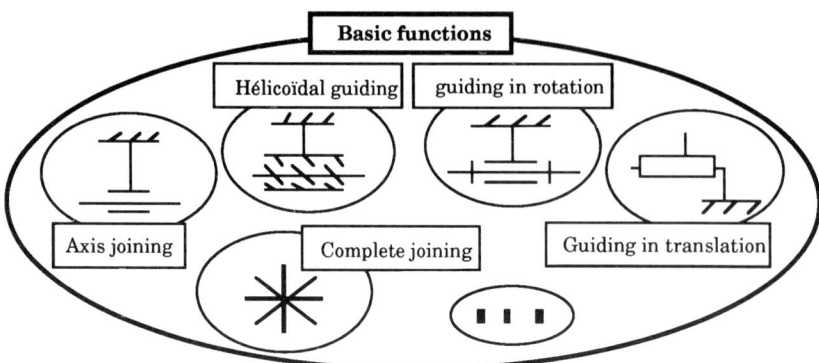

figure 4 : five basic functions

3.1.3 Choice of technological solutions

Based on his experience concerning the problem, the designer chooses one technological solution out of many possibilities to satisfy the previously identified basic functions. He specifies the means by which a basic function is realized. It is important to note that the choice of one technological solution is not definitive (same procedure for the following steps).

In the following, a *technological solution* specifies the chosen means by which a joining is realized and its formulation indicates the authorized or prohibited motion.

As previously illustrated, figure 5 shows various symbols which are associated to some technological solutions. Since no standard exists, these symbols have to be quite imaginative.

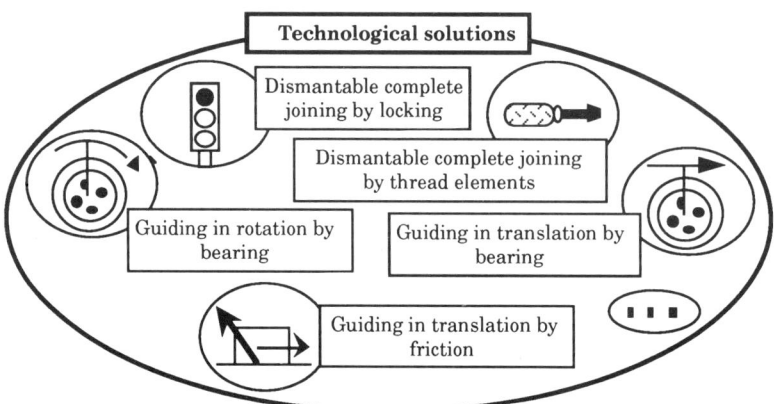

figure 5 : some technological solutions

3.1.4 Choice of technical solutions

Recall the previous step aims to specify the means by which the authorized or prohibited motions are obtained in order to satisfy the considered basic function. Nevertheless, these solutions are always quite abstract. This step aims at changing this by an identification of an arrangement of actual shapes with standard elements (or not) to materialize the considered technological solution.

In the following, a *technical solution* is realized from an arrangement of shapes with standard elements (or not) which authorizes or prohibits a motion satisfying one technological solution.

The following figure 6 illustrates some symbolic representations of technical solutions. These symbols are based on the scheme standards used in the technical drawing.

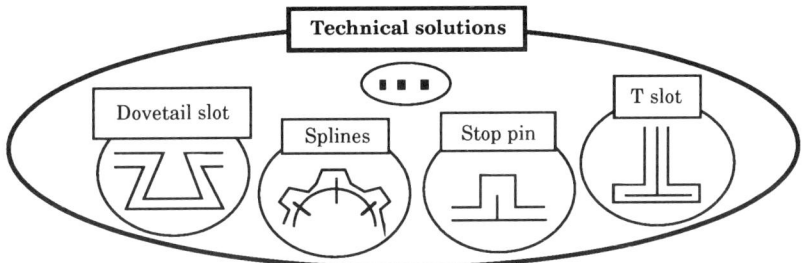

figure 6 : some technical solutions

3.1.5 Evaluation

This step consists in evaluating the realistic dimensions of the technical solution according to the real conditions of use. Recall a technical solution is generally materialized by matting machining shapes. So, this evaluation is local.

In the following, each of the matting functional shapes required on two parts (or more) to materialize one technical solution are named *form features*.

We have just said this step consists in evaluating the dimensions of matting form features within a product. However, the complete design of a product needs to state the dimensions of the functional shapes (form features), the dimensions and tolerances which link the form features (precision features), the global shape of parts within the product, all related to mechanical properties (material features).

At this evaluation step, we only know the various functional faces on the different parts of the product (including their own D&T and their relative spatial orientation). Thus, it is necessary for each part to link the functional faces in order to materialize them within the product. These links are called here *wrapping features*.

Some features we take into account are shown in figure 7.

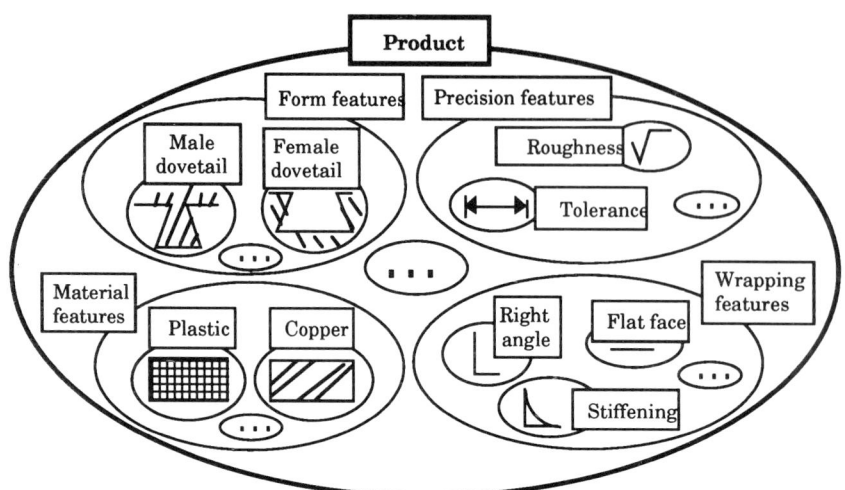

figure 7 : different features in the product

3.2. The take into account of machining

Before formalizing the concept of functionalities, an identification of the minimal set of information which characterize a group of bright faces among others was performed. This aims to move information absolutely necessary to machine a shape from CAM step to CAD step. The result is an identification of the characteristic information of a milled shape or drilled shape and the association of an elementary process plan to each of these information.

Different cases may be considered to take into account machining (figure 8).

1) A form feature. From an identification of its characteristics, a set of elementary manufacturing rules determines its elementary process plan. e.g. :
   IF through-hole, THEN heigth of drilling = heigth of hole+marge,...
This determination may induce a proposition to change dimensions because of the production cost is too high, the required tools are not standard,...

2) Some form features. This determination of elementary process plans offers the possibility to optimize the number of tools. e.g. :
   IF holes with close diameters, THEN using the highest value for all,...
As a matter of fact, a functional value can be increased, but must not be decreased.

3) A complete part.
- Interferences of machining between form features may be identified : a more suitable wrapping feature is indicated. e.g. :
   IF obtuse angle between two milled faces, THEN use a round angle,...
- The process plan of the part may be generated taking into account the available machines in the shopfloor,...

4) The product. Global optimization with Taguchi methods, value analysis,...

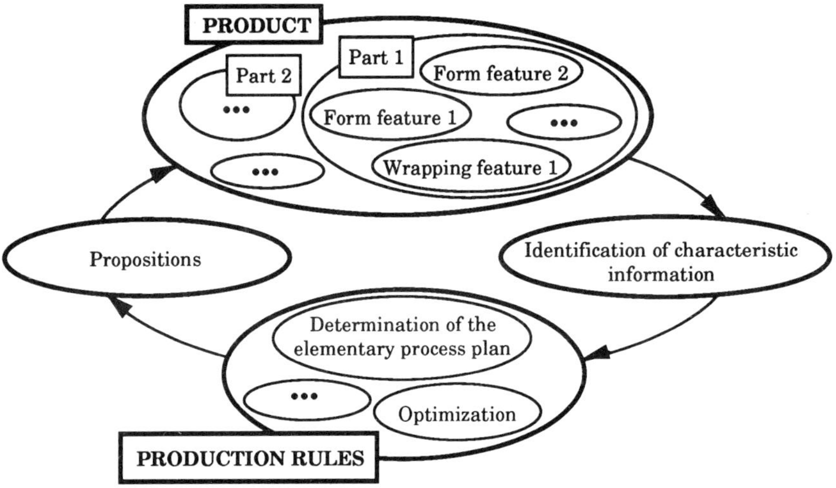

figure 8 : take into account of machining

3.3. Synthesis
The required knowledge for the concept of funtionalities may be classified into three major sets [27] :

- Basic conceptual entities : in this class are regrouped the set of tools, methods and generic theories that govern the basic engineering knowledge. e.g. : geometry, maths, graphs, analysis methods,...

- Basic engineering knowledge : they reflect a knowledge common to all engineers, based on a world-wide know-how. They are not specific to one trade. e.g. : the design rules which allow the designer to calculate the matting form features (bending strength, tensile strength,...), performed an optimization (Taguchi methods,...),....

- Expert engineering knowledge : it represents a knowlege specific to one trade and depend of the know-how in a factory. These rules are based on its past, largely dependant on the available ressources, gained through the experience of the manufacturing engineers,... They are often hard to formalize. e.g. : elementary manufacturing rules which determine the elementary process plan of one form features at a time thanks to an identification of its characteristics, global manufacturing rules which determine the process plan of one part, based on the elementary process plans, the shopfloor,...

3.4. Use of the concept of functionalities
The concept of funtionalities aims to help the user dynamically estimate the dimensions and tolerances of a technical solution suitable to realize a basic

function he identified from the specifications. More precisely, this concept aims to dynamically estimate the dimensions of matting functional shapes (form features), the tolerances (precision features), in relation to mechanical properties (material features) from the specifications.

The use of this concept is shown in figure 9.

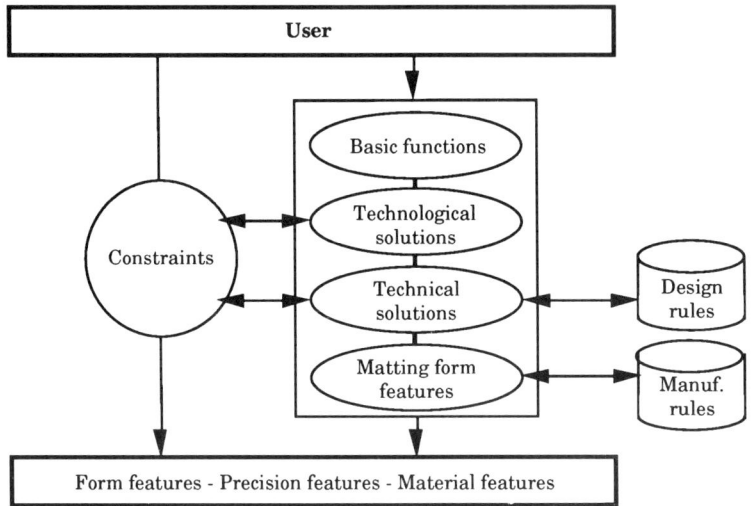

figure 9 : use of the functionalities

Once a technical solution is chosen, the user indicates information which allow the system from the design rules to take into account the real conditions of use of this solution (values of the effort, implantation dimensions, interface dimensions,...). From these values, the matting form features are dimensioned, the precision features are specified and the material features are indicated. If the user is not satisfied with the result, he can estimate another technical solution, another technological solution, or another set of basic functions.

When the user has chosen one technical solution for each basic function, he can link the disjoined functional faces (or form features) of each part, thanks to the wrapping features and so design the shape of the parts within the product.

So, it can be said that this concept is close to the design philosophy, since the user firstly reasons on the functional features "directly" from the basic functions within the product, before designing the global shapes of the parts within the product.

4. EXEMPLE : ASSEMBLY BY THREAD ELEMENTS

Specifications : Design a plate coupling (shown in figure 10) to drive the spindle (1) with the main shaft (4).

This requires to dimension the bolts (3), choose the diameter of support gears (4), compute the length of keys, state the precision of adjustating. In the following, we only detail in the following the choices of the bolts (3).

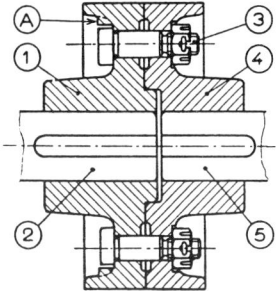

figure 10: exemple of one design by functionalities

Design steps :

Realize a complete immobilization of a part on another means the realization of one complete joining (basic function).

Some possibilities issued from others (shown in figure 11) are offered to the user to materialize it :
- use thread elements
- perform a permanent assembly,
- perform an assembly by adherence and dismantable,...

The first possibility offers the user three choices : use a bolt, a strud or a screw.

When the user has chosen the bolt, the system requires :
- the value of the effort,
- the kind of stress (extension, compression, shearing),
- the minimal number of thread elements and their material.

From these values the design rules are able to :
- compute the minimal diameter of bolts from the classical material resistance rules (basic engineering knowledge),
- determine the nearest standard diameter of bolts and the dimensions of matting form features according to standard tables (expert engineering knowledge)

The result is shown in figure 12.

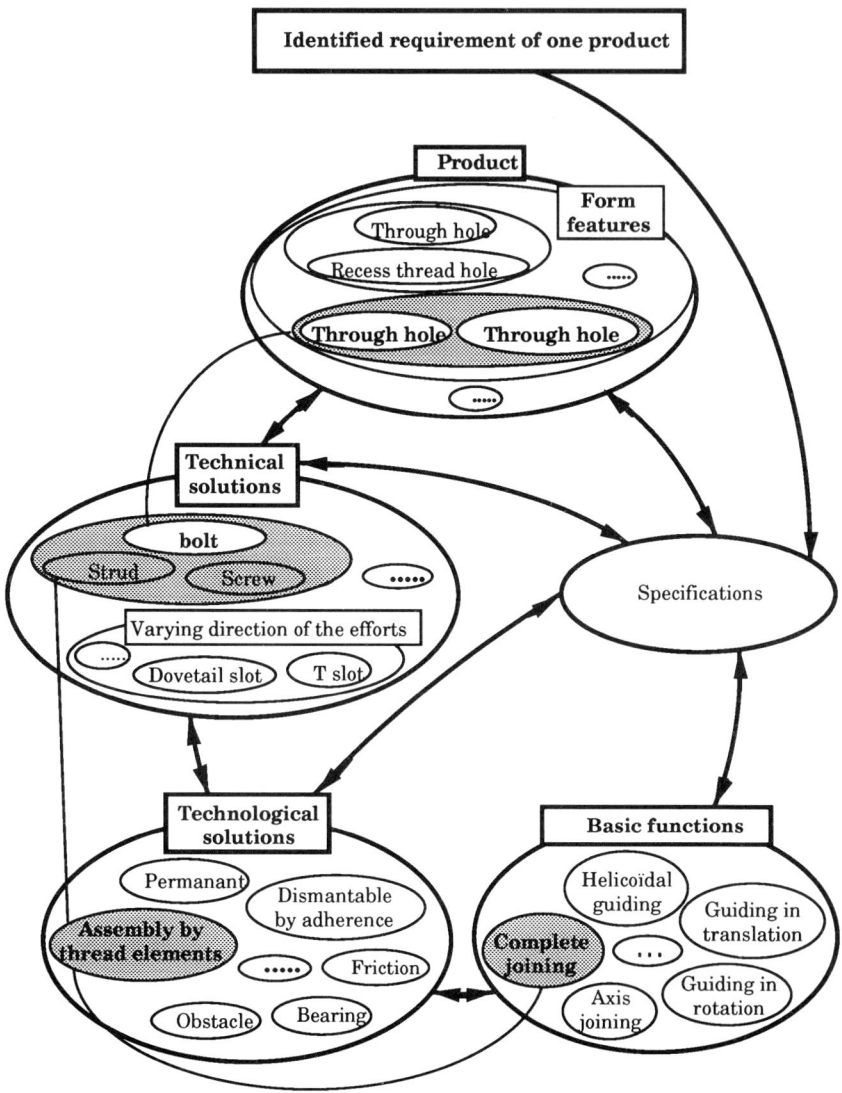

figure 11 : some technical solutions in the case of
an assembly by thread elements (basic function : complete joining)

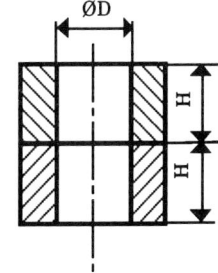

figure 12 : dimensions of the matting form features

The parameters of the holes are :
- through
- diameter "D", height "H"
- material "M"

The elementary process plans are :
- centering hole
- opening drill ØDd*Hd (Dd = d, Hd = H+marge)

5. CONCLUSION

This paper proposes the functionalities as a solution to obtain a continuous informational flow between CAD and CAPP. We have shown here that this continuity is possible thanks to :
- the explicitation of the design intent through the use of basic functions, technological solutions and technical solutions,
- the implicit taking into account of the machining through the association of an elementary process plan to each characteristic parameter of one shape.

The works we are in view are the formalization of the production rules :
- to identify interferences of machining between form features and to propose the use of a more suitable wrapping feature,
- to determine the global process plan from the elementary process plans, taking into account the complete part, the available machines in the shopfloor,...

We are currently implementing on a AI workstation (microExplorer) in a Lisp environment in order to validate this concept. However, it seems that this manner in which a product is designed is a solution to obtain a continuous informational flow between CAD and CAPP.

6. REFERENCES

[1] P. Marks : "What do solids need?". Machine design 12, March 1987
[2] J.J. Shah, M.T. Rogers : "Functional requirements and conceptual design of the feature based modelling system". Computer Aided Engineering Journal, Feb 1988.
[3] J.J. Shah, M.T. Rogers : "Expert form feature modelling shell". Computer Aided Design, Volume 20, N° 9, Nov 1988.
[4] J.J. Shah : "Requirements for support of assembly modelling in a feature based environment". Proposal submitted to CAM-I Geometric Modeling Program, Tempe (AZ), USA, Aug 1989.
[5] A.A.G. Requicha, S.C. Chan : "Representation of Geometric Features, Tolerances and Attributes in Solid Modelers Based on Constructive Geometry". IEEE Journal of Robotics and automation, Volume RA-2, N° 13, Sep 1986.
[6] H. Grabowski, R. Anderl, V. Holland-Letz, B. Pätzold, A. Suhm : "An integrated CAD-CAM-system for product and process modelling". Intl. GI-IFIP Symposium '89, Production Technology Center Berlin, FRG, 1989.
[7] F. Kimura, H. Suzuki : "A uniform approach to dimensioning and tolerancing in product modelling". Computer Application in Production and Engineering ; IFIP 1987.
[8] H. Suzuki, M. Inui, F. Kimura, T. Sata : "A product modelling system for constructing intelligent CAD & CAM systems". Robotics & Computer Integrated Manufacturing, Volume 4, N° 3/4, pp 483-489,1988.
[9] D.C. Anderson, T.C. Chang : "Automated process planning using object oriented feature based design". Intl. GI-IFIP Symposium '89, Production Technology Center Berlin, FRG, 1989.
[10] U. Roy, C.R. Liu : "Feature-based representational scheme of a solid modeller for providing dimensioning and tolerancing information". Robotics and Computer Integrated Manufacturing, Vol 4, N°. 3/4, pp 335-345, 1988.
[11] S. Joshi, T.C. Chang : "Graph-based heuristics for recognition of machined features from a 3D solid model". Computer Aided Design, Vol 20, n° 2, 1988.
[12] M. Cherif, H. Chaouch, E. Cocquebert, R. Maranzana, R. Soenen : "Préparation à la fabrication basée sur un modèle relationnel". Revue Internationale de CFAO. Vol 5 - n° 4/1990. Hermès, Feb 1991.
[13] R. Maranzana : "Intégration des fonctions de conception et de fabrication autour d'une base de données relationnelle". Thèse de doctorat de 3ème cycle, Université de Valenciennes, France, 1988.
[14] D. Deneux, H. Chaouch, E. Cocquebert, R. Soenen : "An homogeneous product working out model for CAD/CAM". 23rd CIRP International seminar on manufacturing systems - Nancy, F, 1991
[15] M.J. Pratt : "A Hybrid Feature-Based Modelling System". Intl. GI-IFIP Symposium '89, Production Technology Center Berlin, FRG, 1989.
[16] J. Nieminen, J. Kanerva, M. Mantyla : "Feature based design of joints". Intl. GI-IFIP Symposium '89, Production Technology Center Berlin, FRG, 1989.

[17] Y. Descotte : "Représentation d'exploitation de connaissances expertes en génération de plans d'actions - Application à la conception automatique de gammes d'usinages". Thèse de troisième cycle - INPG Grenoble, Déc 1981.

[18] J.P. Tsang : "Planification par combinaisons de plans : Application à la génération de gammes d'usinages". Thèse de doctorat, INPG Grenoble, Juil 1987.

[19] P. Durand : "Contribution à la génération et à l'amendement de plans d'actions. Application à la génération automatique de gammes d'usinage dans un contexte CIM". Thèse de docteur ingénieur, INPG Grenoble, Déc 1988.

[20] K. Matsushima, N. Okada, T. Sata : "The integration of CAD and CAM by application of artificial-intelligence techniques". Annals of the CIRP Vol. 30/1/1982.

[21] T. Gupta, B.K. Ghosh : "A survey of expert susyems in manufacturing and process planning". North Holland - Computers in industry 11 pp 195-204, 1988.

[22] M. Henderson, L Taylor : "Capturing the relationship between functionality and features". Unsolicited Proposal to the CAM-I Product Modelling Program, 1991.

[23] J.M. Brun : "Towards integrated product modelling. The contribution of an ESPRIT program". MICAD 91, Vol 1, pp 14-31.

[24] "The modeling of assemblies for design and manufacture". CAM-I report R-88-GM-02, Feb 1988.

[25] F. Kimura, H. Suzuki : "A CAD system for efficient product design based on design intent". Annals of the CIRP Vol 38/1/1989.

[26] F. Knoglu, M. Donath, D. Riley : "Expert system model of the design process". CASA/SME AUTOFACT 5 Conference, Nov. 1985.

[27] T. Sata, F. Kimura, H. Suzuki, T. Fujita : "Designing machine assembly structure using geometric constraints in product modelling". Annals of the CIRP Vol. 34/1/1985.

[28] O. Katai, H. Kawakami, T. Sawaragi, S. Iwai : "Aknowledge acquisition system for conceptual design based on functional and rational explanations of designed objects". 1st international conference on Artificial Intelligence in Design '91, Edinburgh, UK, 1991

Knowledgebased process planning with object oriented implementation

Odd Myklebust

SINTEF Production Engineering
7034 Trondheim, Norway

Abstract
The process planning system described in this paper is a part of ESPRIT II Project 2165, IMPPACT- "Integrated Modelling of Products and Processes using advanced Computer Technologies" The system is a general purpose Manufacturing Decision Support system. capable of Process and Operation planning across a wide range of manufacturing industry.

The process planning system will cover all basic process planning functions, and has available a Modelling tool for exploiting the open information gateway into the part description model. Subsequently, further on the system will be a supervisor or a controller of data in the production preparation phase.

The system will cover the enduser's needs for enabling the activation of operation planning systems of both automated and interactive types. This makes the system both functionally integrated and data integrated.

The process planning system will also be a coordinator in the production preparation phase and make it possible to model product description into manufacturing description.

1. THE PROCESS PLANNING TASK

Process planning is done by a manufacturing engineer. This person needs a detailed knowledge of manufacturing techniques in general, and the facilities available within the particular company. Standard practices are adapted to suit local conditions and customers. This means the engineer needs:
- fundamental manufacturing knowledge
- general manufacturing experience
- company specific manufacturing experience.

Such people are a valuable, though volatile, company asset. It makes sense to provide them as much assistance as possible, and formally capture their local knowledge and data.

2. PROCESS PLANNING SYSTEMS

The majority of process planning being done today is still manual, with computer assistance limited to word processing activities.In the development computer aided systems it has been different generations of concepts, variant planning, early generative systems used FORTRAN or simple decision tables for the logic. Current work moves the elements into an external manufacturing database and gives them more flexibility. The logic is also moved into an external rule-base using more sophisticated decision making or an artificial intelligence (AI) approach.

It is thus easier and quicker to install the system and much of the work can be done by the company's own manufacturing engineers.

The Imppact design is based on this method and offers a powerful, flexible tool with fast payback. Another major benefit is that personal knowledge has been amalgamated into a permanent company manufacturing database.

3. PROCESS PLANNING AND CIM

An examination of the information flow in a design and manufacturing environment shows the central, controlling role of process planning. There are upstream interfaces to design systems, downstream interfaces to manufacturing control systems (scheduling, costing, shop floor control etc) and sideways interfaces to other manufacturing engineering systems (operation planning, NC programming, tool control, raw material stock control etc)

4. CHALLENGES AND APPROACH TO THE PROBLEM

The IMPPACT approach to process planning is to merge the technology available today, that is open information systems using database management tools, with other systems keeping product information. The integration aspect is fundamental. Feature technique is a central method within the IMPPACT project.

The IMPPACT project attacks the integration problem and believes it is possible to turn isolated islands of automation into integrated system by applying and augmenting product information in different preparation stages. The aim is to achieve a quicker and safer method of preparing the production data. When managing the integration aspect, the aim is to achieve an higher level of automation.

Rules are derived from human work. Since a rule is not always expressed precisely, rule validation is necessary. In the IMPPACT project, computer aided process planning is normally understood as an interactive session guided by the process planning program.

A major section of the IMPPACT project is defining an integrated product model. This will contain all the design information in a structured manner.

Because this model is structured it will be possible for the process planning system to interrogate it in an intelligent way. This lets the planning system interpret the information itself, reducing the interactive workload on the planning engineer.

The output from the planning systems is being written to the IMPPACT database. It will thus be available to all other systems within the project. For the external manufacturing control systems we will provide a configurable report writer to supply the desired information in the required format.

5. KNOWLEDGE-BASED PROCESS PLANNING

The process planning task can only be improved through direct access to the product description. This is a *must* if a higher automation level is desired. The IMPPACT feature approach is an important step in this direction. A breakthrough of this principle will be a great achievement and be a basis for later commercial utilization.

This specification covers the principle approach to how a recommended process plan could be generated. Product design evaluation correlated to a feature library keeping descriptions about manufacturing activities, will generate a *recommended* manufacturing process. Some effort is also put into describing how modelling work will appear to the end user, but this is limited to examples about the demonstration activities.

A process planning tool needs to cover any operation type. In that sense this specification merges state-of-the-art technology with new possibilities. The overall goal is to reduce the end user´s efforts in bringing an optimal process plan forward. It is accepted that human reasoning still has to be a central activity and that automated systems within production preparation are not the only goal. The approach is to apply higher automation levels where the end user finds it beneficial. The final decisions is the end user´s. He decides whether decisions should be overruled or not. Activation of knowledge-based modules or reasoning modules can be disabled by the user.

5.1. AI Approach for machine learning

To make process planning systems highly efficient, the user needs the possibility of modifying the decision process of the process planning system and to adapt it to the needs of the manufacturer. It is necessary to reduce the user's interactions in the future. AI supported systems give the possibility of using the user's interactions as examples for learning methods.

5.2. Rule based machine configuration

A typical task of process planning is the selection of machines and clamping devices. In the current state of the art, decision tables and macros are used and a full automation of the decision process is not normally possible. Knowledge based methods however are designed to model dependencies and constraints of the following relationship: work-piece, machine, fixture. In order to infer from rules those of machines, fixtures and machine-fixture configurations which are not suitable for each other and for the given work-piece. In this way, a knowledge based pre-selection of machines and clamping devices relieves the user of checking the possible conflicts between machines and clamping devices and their suitability for the given machining task.

5.3. Manufacturing logic

Manufacturing logic is involves design features and manufacturing activity features. Manufacturing activity features may be considered as a library of what is possible to carry out using current manufacturing equipment. This is part independent knowledge.

When investment is made in new equipment, this knowledge will change. Machine wear and feedback from shop floor also influence this knowledge. This process planning specification addresses the importance of having current updates of this manufacturing logic. The requirement is to have maintenance as a spinoff result from the daily job. Maintenance must not cause delays or be considered as an extra job without any direct benefit to the recipient. If so, the job will most probably not be done. This is the *desirable goal*.

The alternative is to accept that these changes result in a maintenance job. It must then be supported by a powerful modelling tool. In the same manner, when a new product type is being introduced, the modelling of corresponding manufacturing logic will be a part of the job introducing a new product.

5.4. Integration to design

The process planner's need of information to make the process plan automatically or interactively will in general still be the same no matter which data transfer or integration methods are selected. Based on the part identification the process planning system has the address of the part geometry, so it is possible to fetch this information from the design modeler.The use of design geometry information for determination of processes, operations and NC-paths is in IMPPACT mainly served by the feature approach.

5.5. Technological information

Information about tolerances, blank part geometry, surface roughness, surface treatment such as painting, anti-corrosion covering etc. has to be given to the process planning system. All kinds of technical information (non geometry) for the actual part must be stored in a structured way.

If there is less technological information available from the CAD-system to the CAPP-system (Computer Aided Process Planning system), more interactive work has to be done by the process planner.

6. SYSTEM STRUCTURE AND SOFTWARE UNITS.

The block diagram of the process planning system shows seven software units and external interfaces to operation planning systems and downstream manufacturing control systems (e.g.. scheduling).

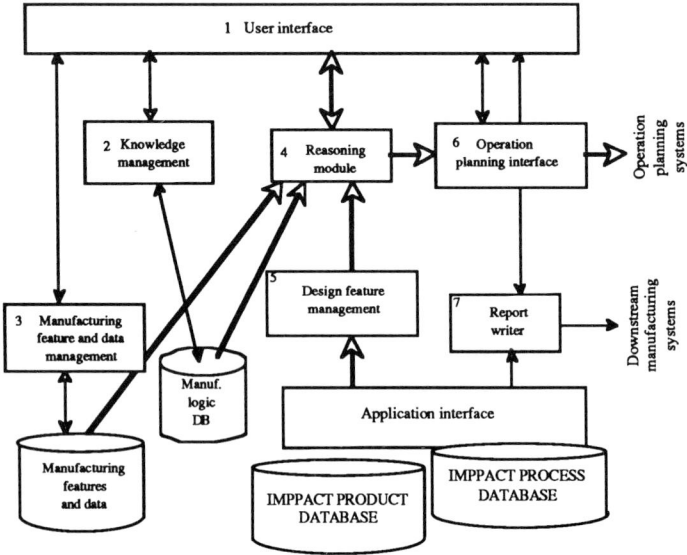

Figure 1: Block diagram of the process planning system

Heavy arrows in figure 1 show data flow for planning an individual part.
Light arrows show maintenance and enquiry data flows.
The process planning systems design is divided into seven numbered modules. These are modules from the development viewpoint, and a split into component subsystems for hardware will be unique to the partner concerned. This description defines the functions that the system will provide in order to support the external, user level facilities that the customer requires.

6.1. User interface and supervisory functions

This unit provides the complete user interface to all functions identified in the User Requirement Specification. It includes screen management, user command parsing and the resulting display of command information. The functionality may be implemented as embedded sub.units or by calling other units.

6.2. Manufacturing knowledge management

This unit manages the user defined, external manufacturing logic. It covers: create, alter, delete and display functions. The display can be text based or a network flowchart. The unit includes printing and plotting interfaces.

The manufacturing knowledge can be managed in two ways. A discreet maintenance session may be started by the user via the user interface. The second way is to dynamically update the knowledge base during an interactive planning session. This means the system can learn from an expert user.

To use the learning mode, the current condition of system parameters are recorded. These provide the input to the decision process. The process planner´s response to these conditions is recorded with them. After a sufficient number of cases have been collected, the data is analyzed and new rules are proposed. These are validated for consistency with existing rules. If the user agrees, the new rules are added to the knowledge base. The level of automation in the decision process will therefore steadily increase.

6.2.1. Rules to generate constraints for machine and clamping devices according to the work-piece description

A rule based system which derives restrictions for the machine and clamping devices from the work-piece description, is integrated. Those descriptive items of the work-piece, which will be essential for the selection of a machine and clamping device. The rules handle those items in their premises, and the constraints for the machines and clamping devices in their conclusions.

6.2.2. Rules to configure machines and clamping devices

A rule based system will generate all possible configurations of a given set of machine and clamping devices. The system will consider those descriptive items of the machines and clamping devices which are essential for the configuration. The rules handle those items in their premises, and methods for configuration in their conclusion. A function will check the constraint satisfaction for the machine and clamping device configurations and choose the matching ones as possible configurations for the machining task. Finally, the user can choose a machine and clamping device configuration out of a sequence, which is ordered by a function for evaluation with user input of the ordering criteria.

6.3. Manufacturing feature and data management

This unit provides the standard create, alter, delete and display functions for all the items in the manufacturing database. The most important of these are the manufacturing activity features.

6.4. Reasoning module

The reasoning module is a standard but extendible unit to determine which items are required in the process plan for a given part. The reasoning module accesses the manufacturing logic database. This logic is controlled by the part description and manufacturing features available. It generates items into the process planning database, including time and cost information.

6.5. Design feature interface

This is an interface unit which will scan the part description for known elements. The identified elements will be grouped or classified, and then presented as current knowledge about the part being planned. The consumers of this knowledge are both the user and the reasoning module.

6.6. Operation planning interface

This unit scans the part description and compares it to its knowledge about which specialized external operation planning systems are available. The module can both identify this on its own (a parallel process to the reasoning module) with confirmation from the user, if required, or be activated by the process planner.

6.7. Report writer

This unit is the primary method for humans and non IMPPACT systems to obtain information from the product planning database. It allows a user to define the shape and appearance of the form, and to specify which variable information is to go in each field.

7. USE OF FEATURES

Features are seen as the integration between the design and planning system. The manufacturing oriented features will be used to reduce or eliminate interactive user input.
Design information from external CAD systems may not contain a feature representation. In these circumstances a process of feature-recognition is required. It is likely that this will involve user input and graphical manipulation of the geometry. From a process planning point of view we believe it is more appropriate to associate this task with the design system. The process planning system will then not have a feature-recognition capability.

7.1. Scenario of feature based process planning

When the CAD work-piece model has been approved and released for production, the process planning system accesses the CAD work-piece model and looks for applicable features. The design features are grouped according to feature classification and prepared as input to the manufacturing logic.

According to the feature occurrences, manufacturing logic is evaluated. A reasoning module is an active task working on manufacturing logic which accesses the extracted design features and available manufacturing activity features. When logical connections are found, recommended elements of the process plan or input to other manufacturing logic are generated. The user is consulted when the logic meets "equal" branches or other undefined expressions. Manufacturing logic which can access MRP information is necessary. The reasoning process argues with the user, and the user always has the last word.

If the current part has design features which trigger clamping logic, the user notice that *assistance is available*. If he chooses to enable the clamping selection system, that task is executed and the result included into the process description.

Manufacturing logic is currently being maintained through disagreements between proposals from the reasoning module and the user. The logic was initially built up when the product type was introduced. The extent of the manufacturing logic is dependant on feature occurrences. As the number of features grows, new relations are possible.

7.2. Feature classification, libraries

Features are used in several disciplines which is how they can solve the integration problem. A feature occurrence may hold both a design view and a process view. When a "consumer" asks for information, the required view has to be supplied. Then the appropriate presentation is chosen, according to whether the information is going to be used by a reasoning module or a human.
There are two main groups of features:
- Design features
- Manufacturing features.

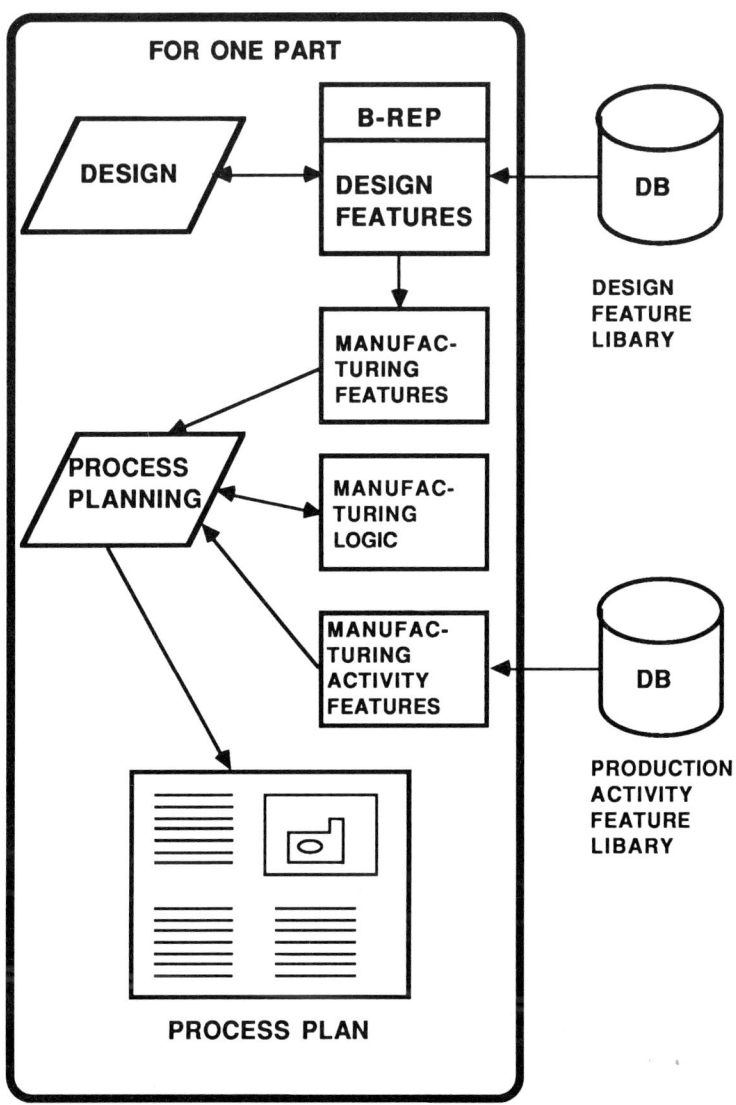

Figure 2: Features used in a process planning environment

The IMPPACT project has been doing fundamental research work on the feature approach to the integration problem. The work has to start very close to the design task where the product is being described. Definition of form features is therefore given most attention. The further use of form features is continuing within the detailed operation planning and NC programming tasks. Process planning uses also form features to classify the product type and through this draw different conclusions. By using the management system to create new features, then the range of exploitation is in principle unlimited.

8. THE SOFTWARE IMPLEMENTATION WITH XPLAIN, OBJECT ORIENTED DATA STORAGE.

XPlain is a software development tool available on UNIX workstations using the X Window System. XPlain has a rule based language to describe the interface between the user and the application or databases. XPlain Designer is used to design the user interface including graphics. Close integration to R-DBMS with SQL query statements accepted in the rule language. Xplain also support the Object oriented database system prototype developed in Imppact.This ODBMS system are used for all storage in Imppact product database. The database schema input is based on EXPRESS which is used a the overall modelling language in Imppact.

The EXPRESS language is a powerful language which allows us to specify data-structures very accurately. It allows us to define entities with their attributes and relations with other entities. But it has also more powerful abilities in the form of Pascal-like functions, Rules, Where -clauses, etc.

XPlain combines features from advanced user interface management systems (UIMS), expert systems, spreadsheet systems, sketching tools and database systems. The layout description of the user interface is done by sketching interactive forms and the functionality is defined in expert system like rules.XPlain is not a code generating system nor an interpretative system. When your XPlain application is running, you may step into the rule editor, change rules or add rules. Then you may step back into your running application just by pushing a function button. The compilation is done when exiting the rule editor, no system generation is necessary and the performance is high.XPlain has inter process communication available. It is possible to pipe results from an external running process into a form sheet field.

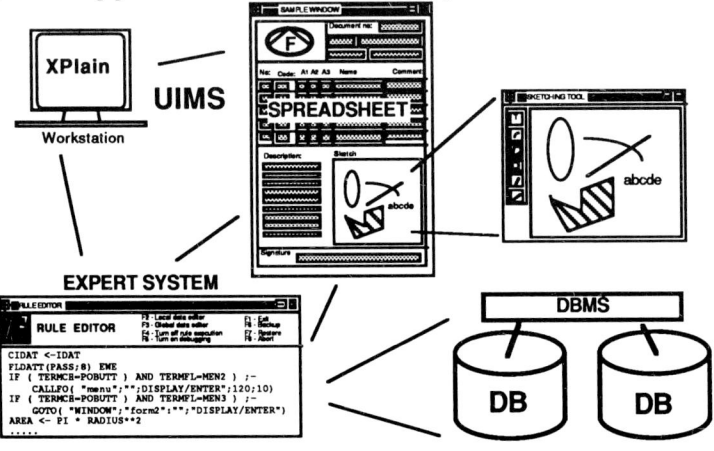

Figure 3: XPlain joins features from different types of programming tools

9. REFERENCES

1. Romstad Arvid   User requirement specification for the process planning supervisor, version 2, SINTEF January 1990.

2. Roger Crawford User requirement specification for the Locam, process planning system, Revision 11, PAFEC, March 1990.

3. ESPRIT-IMPPACT, WP3: Deliverable D301, Functional requirements for process modelling, 14 August 1989.Document - #D.WP3-CEC.#3000/001

4. Leira Jarle   Software development handbook, Version 3 SINTEF, March 1990, Document - # T.SIN-AIP.1600/001 (Appendix to deliverable D111).

5. Wolfgang Teichmann   Learning modules, AI Techniques in process and operation planning Krupp Forschung 1990 (Document)

AN INTEGRATED SOFTWARE SYSTEM SUPPORTING A MACHINING CELL IN MECHANICAL ENGINEERING

Prof. Nicola Todorov, Ph.D., Mech.Eng. and Iossif Levi, Mech.Eng.

CAD/CAM Center of Machine Tools Institute, P.O. Box 107, 1000 Sofia, Bulgaria

The Group technology (GT) is a manufacturing philosophy which main idea is to make use of the similarity and repeatability of different activities. It is a philosophy with a broad base and application opportunities affecting all areas of a production organization.

A specific application of the GT can be the production in machining cells (MC) - cellular manufacturing (CM). CM makes use of the combined machining technology of similar parts (part families) on different machine groups.

The CM goal is to reduce the setup times using the part-family tooling and sequencing, and flow times reducing setup, transport and waiting times. Thus inventory stocks are reduced as well as market response times. In addition, the MC have social aspects which predispose for a team work with a natural high quality production motivation.

As a result of this, CM has been recognized as a promising strategy for the rejuvenation of outdated and unproductive plants.

The new technologies often both support and mandate a CM approach. Support means involvement of information technologies, such as computerized part coding schemes which help in part family identification for cell production. From the other point of view the demand for CM derives from the application of robotics and other forms of mechanized/automated material handling systems as well as from the desire to build closely related manufacturing systems with low throughput times. Efficient use of such systems and technologies requires a cell-structure approach to manufacturing.

There is no doubt that CM represents a major technological innovation to most organizations. CM, therefore, warrants particular attention from researchers.

In this paper the developing project of a System for supporting and management of manufacturing cells is discussed.

The main goal of the project is making use of the existing and new information technologies in order to make available the most appropriate opportunities and convenience for working in CM. The main idea is to create a frame in which an information environment supporting all the activities in CM. It will be possible to make decisions on a lower level and the available computer tools will give the opportunity to the users to check variants using the data from all the stages of the lead cycle and to test their ideas in choosing the best variant for action. This will lead not only to a better effectiveness but also will raise the satisfaction of the people from their work.

The decision making in the execution point will result in a reduction of the information cycles, in a less data flow, in a smaller reaction time to changes or disturbances and consequent changes in the day picture of the manufacturing process as a whole.

As a result from the project an IT product will be created. It will cover the whole range of manufacturing activities in the manufacturing cell. With its help not only the production team rights and responsibilities will be enhanced but also the volume of the data flow as well as the distance travelled by it will be reduced in comparison with the contemporary conventional management systems.

The system consists of hardware and software tools and allows the following opportunities:

.information support of the process planning in CM.
.simulation of the manufacturing process as a time function
.cluster analysis of the part variety
.information support of the decision making in the case of disturbances in CM
.dispatching of the manufacturing process
.supporting of DNC.

The structure of the system is represented on figure 1 .

The PRODUCTION PLANNING module covers the activities connected with the item structure description and the production orders. This module supports production plans variants.

The PROCESS PLANNING module covers the process planning activities in CM. Here are included the creating of route technologies based on the variant approach and the related documentation: route plan, control plan and equipment plan. For the work plans generation a customized expert system is used.

The STATIC SCHEDULING module covers the activities for creating of month, week production plans and scheduling. Here there are possibilities for testing of production plans variants through simulating of the production process. Using the cluster analysis methods a part analysis is made aiming at their selection for manufacturing in CM or at the design process.

The SCHEDULING AND CONTROL module covers the activities scheduling, dispatching and reporting. Through this module the connection to DNC and to the information support of the machine tools is made.

Modules serving the tooling, the transport and finished production warehouse can be joined.

In this system approach there are two specific moments. The first one is that everything starts from the two main sources of engineering information - the item structure and the orders and on the other hand the route technology. The control functions are secondary - they only handle the information but do not create new information. These three parts of (item structure and orders, route technology and control) form the base for the tools of the manager of the manufacturing cell. The processes handling the DNC connection and their control are background processes which work directly with the data base. Systems like „tool room", „transport control",

CELL SUPPORT SYSTEM

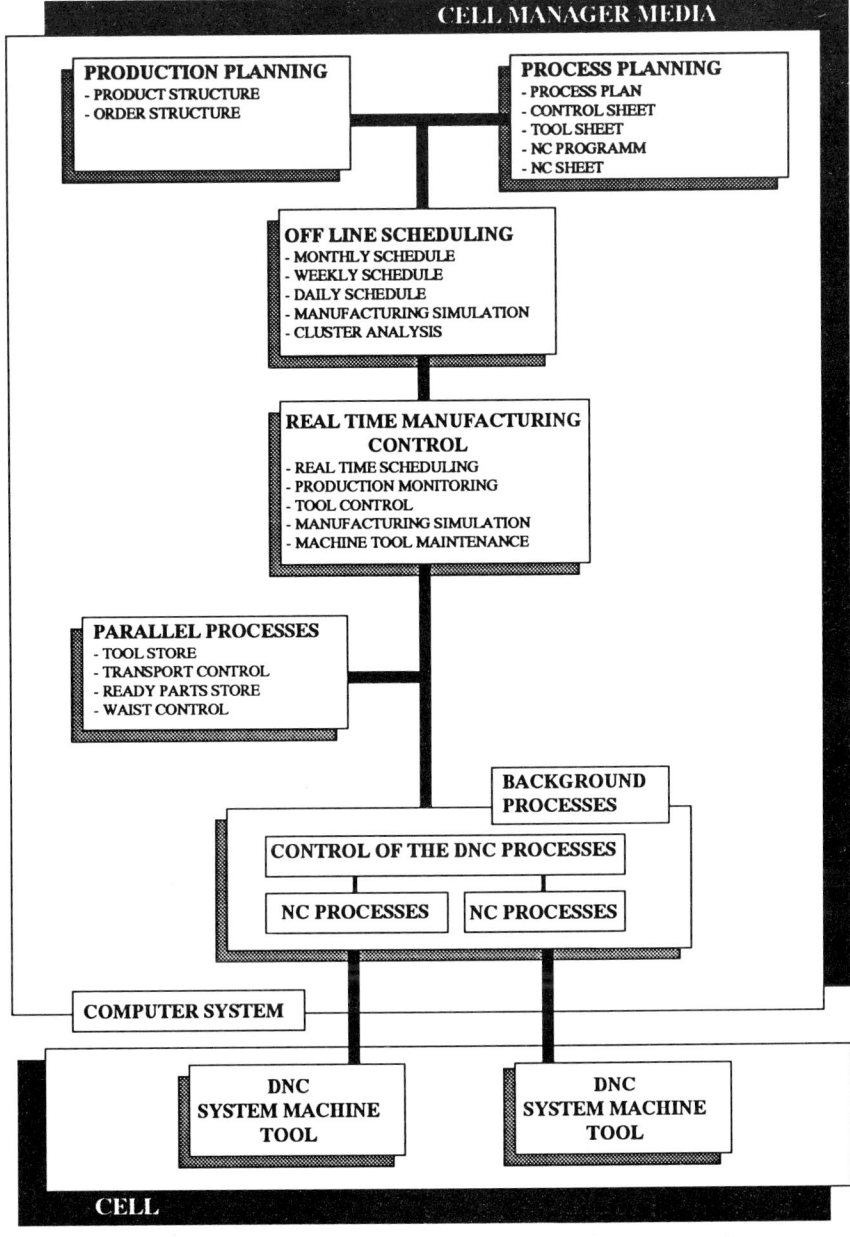

„finished parts warehouse" are optional. They are functioning as parallel processes. They are also connected directly with the data base.

The second specific moment in this approach is the use of an integrated single data base for all the modules in the system. We must stress here that the term of integration differs from the term of interfacing. One difference between interfacing and integration is that interfacing must be addressed at the task level. In other words, it would be too late to integrate a task when its sub-results (such as design and manufacturing specification or decisions for process and operation planning) are already decided separately. Currently a few researchers indeed provide some promising approaches in terms of integration CAD/CAM. However the results carried out are still far from the real integration production. The current approaches are mainly trying to interface various separated activities at the design, manufacturing and planning phase. Each of the phases has its own stand-alone relational database and corresponding Data Base Management Systems (DBMS). There are great difficulties to interface all these separate activities because of technical problems of software and hardware. In order to realize the final integrated production it is probably an ideal approach to integrate all the information involved in producing a product into a Single Data Base (SBD) instead of interfacing all the stand-alone databases. The integrated SDB may include all the data from design, analysis, drafting, process planning, and NC tool paths as well as the project of planning, bill of materials, and production scheduling information, post-process planning information, etc.

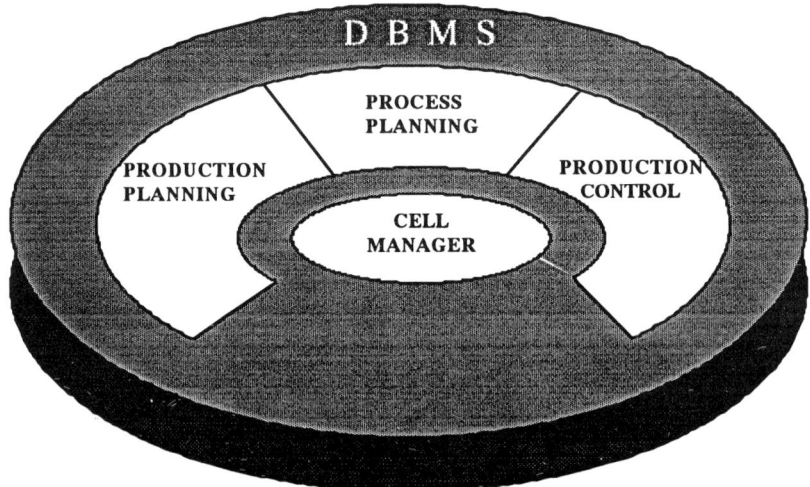

All modules in the described structure are working with one and the same data base and there is no need in creating of special connections between them.

The specific features of the cell are reflected as data in the data base. If a user wants to change the cell structure he must change only the data in the data base. On the other hand the users with different manufacturing cells have similar manufacturing cell supporting systems but with different data.

The system makes easier to take the advantages of the GT philosophy in the cell manufacturing.

It gives new opportunities to the engineers - to work with a system which has a real time connection with the machines in the shopfloor. Thus the process planning, NC programming and the scheduling using all the production information is made possible. And also the benefit from the masters and operators skills and experience is used for solving the contradictions arising from the differences in the approaches in the process planning department and the shopfloor.

The software part of the system can be used in the machining cell development process which makes easy the transition to the new work organization. It serves as tool for part families and machine groups forming.

This project is based on already existing process planning software which has several industrial installations.

PROCESS PLANNING FOR COMPLEX MACHINING

CHAIRMAN: R. SOENEN
UNIVERSITY OF VALENCIENNES, FRANCE

CAPP based on advanced modelling techniques

H.-K. Tönshoff[a], M. Becker[a] and J. Kreutzfeldt[a]

[a]Institute for Production Engineering and Machine Tools, University of Hannover, Schloßwender Str. 5, D-3000 Hannover 1, Germany

Abstract

This paper presents results of research in the area of Computer-Aided Process Planning (CAPP) for discrete part manufacturing. A new approach to generative process planning is described. The delineated concept makes great demands on the modelling of intermediate workpiece states during the manufacturing process. Nevertheless, the approach has already been successfully applied in industrial projects for workpiece families whose modelling requirements are relatively elementary. Recent progress in the area of feature-based modelling, leading to a new generation of CAD-modellers, will allow the investigated approach to be applied to a broader workpiece spectrum in mechanical engineering. This paper presents not only the planning approach itself but relates it to current developments in geometric modelling.

1. SITUATION OF CAPP

CAPP represents the decisive link between CAD and the production process. Because of its important role as a functional module, CAPP is now recognized as an essential component for the realization of a CIM scenario [2, 24]. The operation sequences and the associated technological parameters listed in the process plans embody the overall knowledge about manufacturing processes available within a company.

The improvements in CAD and CAM techniques and their intensive and widespread usage in the industrial environment have generated an urgent need to develop new methods for CAPP. One of the major goals is the reduction of the time needed to create a particular process plan. The necessity of troughput-time reduction in the process planning department becomes clear when considering that working without an ordinary process plan leads to an inefficient and cost intensive production. Investigations have shown that 20% of all orders in small and medium size enterprises are manufactured without a process plan, or not according to the previously generated process plan.

There are further factors affecting the increasing demand for research activities in the field of CAPP. One originates from the fact that different process planners generate different process plans for a single workpiece, according to their individual manufacturing knowledge and experience. Consequently, the results of process planning are not as reproducible as necessary. This has a negative effect on the reliability of the offers a company can make, because the cost estimations are based in general on time data included in the process plan. Therefore, the generation of operation sequences and their time data must be standardized to ensure that similar parts lead to similar process plans. This could be done by providing standardized information data bases and automated tools for the generation of process plans.

From a survey of research activities in the area of CAPP and the related system implementations one can clearly recognize the major difficulties all systems encounter [33]. One problem is the modelling and representation of knowledge concerned with manufacturing processes, which is difficult to model mathematically and program with common procedural languages. Another major problem is the automatic acquisition of the workpiece geometry. Many of the related studies propagate feature recognition from a 3D geometric workpiece model as a sufficient CAD interface. But up to now their capabilities are strongly limited.

2. APPROACHES TO CAPP

The first steps towards computerisation in the area of process planning resulted in program systems for the storage and retrieval of process planning data. The process planner is provided with text processing functions and a support for the administration of generated plans.

In *similarity planning*, the computer searches and retrieves a similar workpiece and its process plan. The planner changes the plan according to the current workpiece. Similarity planning presumes a classification of the workpiece spectrum. The classification may be achieved firstly on the basis of design information. In this case the workpiece spectrum is analysed, either manually or automatically, by investigating the CAD database of the company. The latter solution can hardly be realized if the CAD representation is not feature-based. Secondly, a classification may be carried out on the basis of process planning information. In this case, existing process plans are investigated in order to identify either a similarity in terms of workpiece characteristics such as weight, length to diameter ratio, etc, or whether the correspondence in the operation sequence can be used to determine a workpiece group. Cluster analysis was applied successfully for the analysis of both design and process planning information in order to identify corresponding characteristics of different workpieces [9, 10].

In *variant planning*, the computer generates the complete process plan on the basis of a standard program defined by the process planner for a complex workpiece. During the planning task the process planner describes the given workpiece by setting parameters predefined in the variant program.

In *semi-automatic generative process planning*, the planner determines the operations sequence and tool paths interactively. The computer calculates technological-, cost- and time data. Such systems presume a technology database (eg. cutting speed data), the workpiece description and a time and cost database (operation times, set-up times).

Most of the CAPP systems marketed today aim at semi-automatic generative process planning. These systems can be characterized by the degree to which they fulfil the following two key requirements. Firstly, the provision of necessary planning data in terms of technology, times and costs, and secondly, the support of the planning task itself. Many of the commercially available systems sufficiently fulfil one of the two functions but have drawbacks with regard to the other one. The most common support of the planning task consists of a decision table system which the user has to configure himself.

In the most challenging approach - *automatic generative process planning* - the computer generates process plans for a heterogeneous workpiece spectrum. The process planner makes no complex decisions, or only very few, which are difficult to automate. Considerable research efforts have been addressing this planning problem, resulting in a number of research prototypes eg. PART [36], AVOGEN [1], EXCAP [5], GARI [6], PROPLAN [21] and [15]. Only very few and limited automatic generative systems have been applied in industry so far: eg. GUMMEX [17], GENPLAN [35] and CHAINPLANNER [32].

In Section 4 a research prototype is presented. Beforehand, the specific requirements of automatic generative process planning are outlined.

3. REQUIREMENTS FOR AUTOMATED GENERATIVE PROCESS PLANNING

3.1 Modelling techniques

There are two areas where geometric modelling techniques have become important for generative process planning purposes. The first refers to the design stage where the workpiece shape is defined. AI-based generative process planning requires a workpiece representation which is sufficient to retrieve process planning information in an automatic way. The second area of modelling requirements becomes apparent if we consider the state-space approach for the planning task (see section 3.2). This planning concept presumes that the modelling component of a generative process planning system is able to handle the intermediate workpiece states which occur during the planning procedure.

Automatic CAPP data acquisition

First of all an overview about the geometric modelling techniques with respect to automatic data retrieval for process planning purposes shall be given.

In 2-D CAD systems the part shape is modelled by means of simple lines and curves. For the process planning task however, the information about manufacturing requirements, added to the CAD-drawing by the designer, is not available in a suitable way. In the CAD-database, technological information is represented only as simple text elements not related to a specific part contour. Consequently the procedure of data acquisition for CAPP purposes becomes very ineffective and also creates redundancy problems.

The next step towards advanced CAD-modelling was the introduction of 3D geometric shape representation. Using this technique, the workpieces can be described with the aid of geometric elements like 3D lines, curves, surfaces or solids. In spite of the increasing modelling capabilities that 3D systems incorporate, it has been shown that designers use the 2D drafting part of such systems most of the time: advanced 3D modellers degenerate to simple electronic drawing boards. This is because modelling in three dimensions becomes ineffective if a designer can only work with abstract geometric primitives. For this reason, further approaches to CAD-CAPP integration could not benefit from the developments in geometric modelling [23]. To get a real increase of productivity using 3D modelling systems the forthcoming CAD-user interfaces and internal geometric data structures must support product modelling functions.

The use of features for the workpiece description seems to be the most promising approach to product modelling today [4]. CAD-systems have to provide a feature-oriented user interface and some systems do this already (ProEngineer, SiGraph CAD SPAN). In general, there are three different feature classes. In the field of *construction* methodology one can identify features to fulfil functional product requirements. During the product design stage, special *design* features could be utilized to reduce the necessary effort for workpiece modelling in three dimensions. For purposes of CAPP, these design features have to be mapped to *production* features which represent the required manufacturing operations. A design feature could also be a production feature, but this is the exception. Usually, complex relations have to be evaluated to derive production feature information from design features.

Many studies concerning design feature catalogues made it apparent that there are different aspects which determine the definition of feature characteristics [8, 20]. Up to now, all known design feature sets are application dependent. A generalized feature catalogue will hardly be achieved and is probably even not expedient [25]. The use of configurable design features seems to be the solution for generating application dependent feature sets. The configuration tool will be used to create complex interrelations within feature structures with the use of a set of basic feature prototypes.

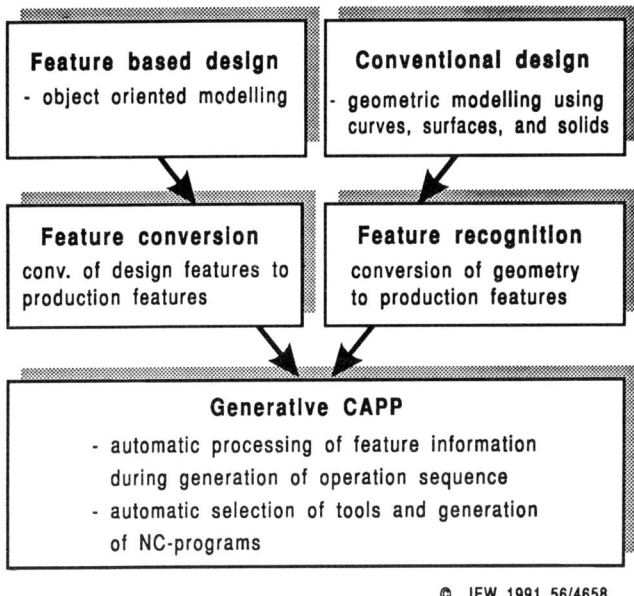

Figure 1. Feature based design vs. feature recognition

Due to the internal data structure of the CAD model there are three conceivable approaches to retrieve feature oriented production information for process planning purposes in an automatic way [11, 16, 18, 19, 27]. If the workpiece model is created with simple 3D geometric elements, i.e. surfaces or solids, a recognition process could be applied to identify production features. If the designer uses production features only to create the workpiece model then the created feature information could be transferred directly to a CAPP-system. If the designer uses design features the previously mentioned feature mapping process has to be applied to retrieve production feature information. The latter approach seems promising as designers prefer design features to fulfil the product requirements. Figure 1 shows the abovementioned procedures.

Modelling requirements within the state-space approach

The planning process using the state-space approach is principally categorized as a guided search within the possible states of a workpiece from the raw material to the finished part or vice versa. During the search process the evaluation of different planning steps in parallel is often necessary. These resulting parallel branches in the manufacturing sequence have to be justified i.e. according to the minimum number of necessary operations to determine an optimal process plan.

Each possible intermediate workpiece state in an individual planning branch has to be simultaneously recorded in the geometric workpiece model to make sure that operators are applicable (see section 3.2).

This leads to the demand that a geometric modeller, useful for this specific type of planning strategy, is not only able to perform geometric operations on the workpiece solid model but also includes an administration mechanism to manage different states of the model. It must be possibly to *"roll back"* a model to a state created earlier or *"roll forward"* to a previously reached state. The functionality of roll back and roll forward has to be available both between states in a linear sequence and between branched arrangement of states.

Conventional solid modelling systems included in most CAD-systems up to now do not accomplish the prerequisite of a built-in state administration facility. Consequently automatic process planning systems based on the state-space approach have to provide their own management mechanism for the intermediate workpiece states.

Recently a new object-oriented geometric modeller (ACIS), implemented in C++, was presented. This promises to make the organization of intermediate states easier. In view of the capabilities provided by the implemented flexible *states management* function it can be anticipated that this gemetric modeller could certainly be useful in further developments of automatic generative process planning systems based on the state-space approach.

3.2 Planning strategy

With the introduction of AI techniques to the field of mechanical engineering, a number of rule-based process planning prototypes were developed [13, 14, 22, 26]. It was anticipated that these systems would allow rules to be expressed independently while interactions between rules would be handled by the system, implicitly taking into account the rule's content. Although significant research effort was spent, very few systems were implemented in industry. The first reason was the abovementioned lack of sufficient modelling techniques. The generation of a suitable workpiece description, especially for prismatic workpieces, often consumed more time than was gained by automatic plan generation. Secondly, the maintenance of the rule base caused significant difficulties. It became apparent that, even with the help of AI techniques, it remains difficult to keep the rule base consistent. This is illustrated in figures 2 and 3.

In the right column, the pictures show a list of operations. From this list, operations are selected to appear in the process plan to be generated. They appear, if the condition in the left column holds true. The order of the selected operations always remains the same as in the right column. This is not the only way to determine the position of the operations in the plan; the important characteristic of this planning method is, however, that the rules depend on the finished part description.

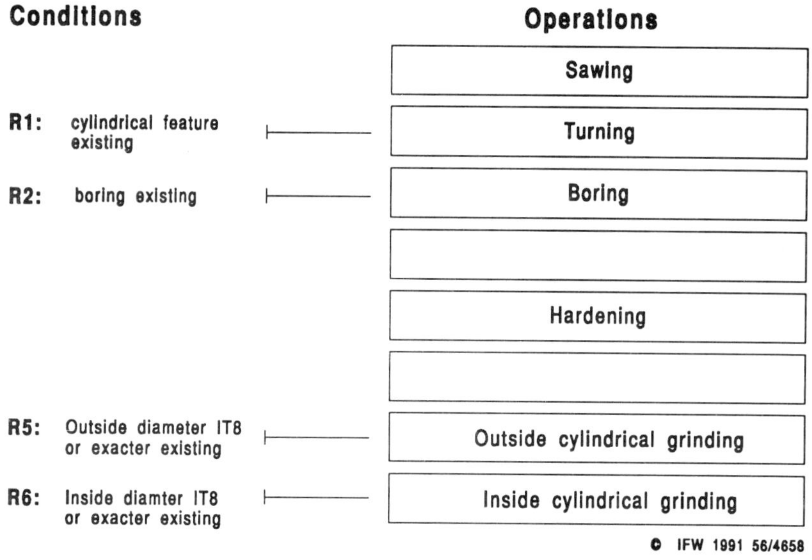

Conditions			Operations
			Sawing
R1:	cylindrical feature existing	⊢—————	Turning
R2:	boring existing	⊢—————	Boring
			Hardening
R5:	Outside diameter IT8 or exacter existing	⊢—————	Outside cylindrical grinding
R6:	Inside diamter IT8 or exacter existing	⊢—————	Inside cylindrical grinding

© IFW 1991 56/4658

Figure 2. Initial sample rule base

The first figure shows a small "rule system" whose rules R1, R2, R5 and R6 determine the necessary manufacturing operations depending on constraints derived from the workpiece representation. Figure 3 illustrates the consequences if two changes are included.

The first change reflects the possibility of increasing production efficiency by the use of a turning centre if both turning and drilling features, having a specific diameter, are available. More abstractly expressed, the combination of two individual features determines the choice of a specific manufacturing operation. Note the necessary adaptation of R1 and R2.

The second change is necessitated by the inclusion of a manufacturing guideline stating that inside grinding shall be performed before outside grinding, if a mandrel for the inside boring diameter is available. This change requires a modification of R6.

Again more abstractly: a given workpiece feature not only determines the choice of a specific operation (grinding with mandrel) but may also cause the exclusion of another operation. The definition of one rule (R6) has to be adapted to take into account the possible previous application of another rule (R4).

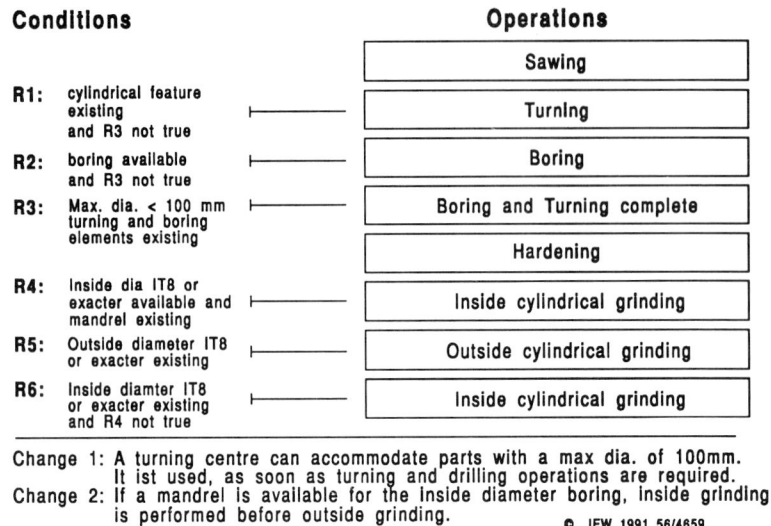

Figure 3. Updated sample rule base

This example emphasises reasons for the interdependence of planning rules. This is caused by the interference of manufacturing constraints which are determined

- a) initially by the composition of the workpiece feature set and
- b) during the planning process by previous decisions which constrain the further application of other rules.

Similarly, one can formulate two requirements which have to be fulfilled in a purely rule-based system in order to allow efficient planning of a workpiece spectrum:

- a) The rules must take into account all possible feature combinations.
- b) The definition of the IF-part of each rule must consider the possibility that other rules (on which this one is dependent) may have been applied previously.

Investigations have shown that, using the method discussed above, both requirements can be fulfilled to such a degree that the rule base can be maintained consistent only for fairly simple workpieces. For realistic planning it can be concluded that rules should be formulated as independently of each other as possible.

The only solution to this problem is for all intermediate workpiece states (during the manufacturing stages) to be modelled. Rules must be replaced by "operators" which test their applicability to the workpiece state. An operator has to perform three tasks:

- a) Test its own applicability to the actual intermediate workpiece state, and, if applicable
- b) update the workpiece state to the new intermediate state, thereby reflecting the real-world manufacturing operation, and
- c) note the textual description of the operation in the plan.

During plan generation, every workpiece state is analysed to determine which operator can be carried out next. If no operators are found a backtracking mechanism is carried out.

Coincidentally, a similar approach was the cornerstone of AI-planning research since the late sixties, also known as the state-space planning approach [12]. Two object classes are of major importance: situations and operators. Situations are models of states of the "world" during different planning stages. Operators are activities which transform one situation into another. The plan generation is a search for a sequence of operators which transforms the initial state to a goal state which satisfies the goal criteria.

The application of the state-space approach on the process planning problem has a vital structuring effect on the resulting process planning system. This ensures that software is well structured and easy to maintain [32].

AI research has concentrated primarily on domain independent search strategies. Until the present, these have proved insufficient to solve the problems encountered during process planning. For this reason, the search needs to be constrained by the use of domain-specific knowledge. This knowledge is provided by the process planner in the form of heuristics. The ultimate goal is to restrict the search process in such a way that the correct plan is achieved without the need for backtracking. Three main models can be derived (figure 4).

The workpiece model covers the feature-based workpiece representation. It must be capable of modelling not only the finished workpiece but also the intermediate states in order to allow backtracking if the application of an operator fails. (see section 3.1).

The workshop model represents the properties of the machines in the workshop. It consists of the total set of operators, ordered according to the machines they belong to. If an operator is linked to a number of similar machines, a further structuring into instances of a common class is performed.

The planner model represents the heuristic knowledge of the process planner. Depending on the considered workpiece spectrum, it may contain several knowledge bases, eg. process selection, operator selection, machine tool selection, fixture selection, etc. According to the appropriate rules of the planner model, the operators to be applied next are selected. If the search for a suitable operation sequence fails, a backtracking process is carried out.

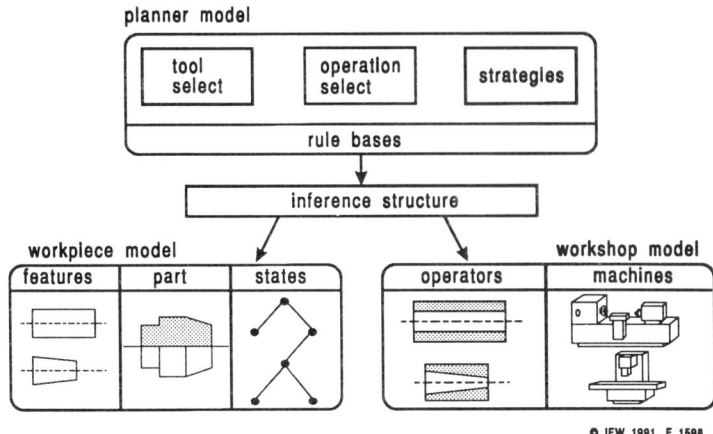

Figure 4. Architecture of the *AVOGEN* process planning system

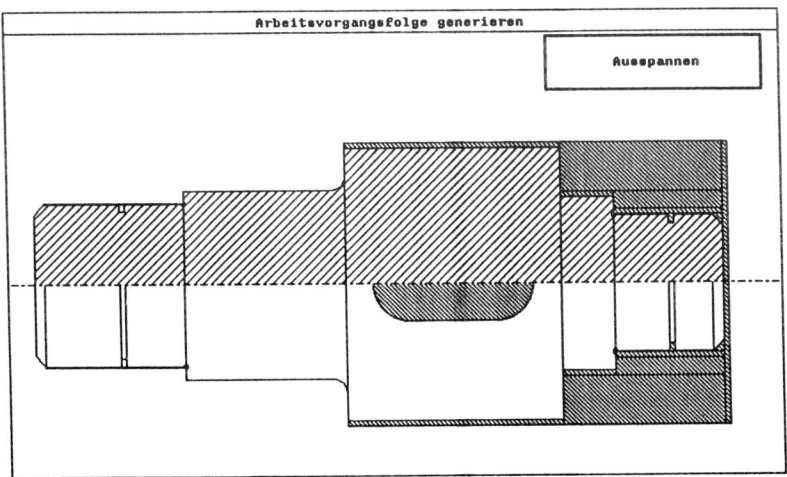

Figure 5. Overview of applied operators

4. REALIZED PROTOTYPES

The rotational part planner AVOGEN allows the generation of company specific process plans for rotational workpieces including several prismatic features such as grooves or keyways. The concepts described in section 3.2 have been implemented. Figure 5 gives an overview of the applied operators during the operation planning of the right-hand-side of the workpiece.

As shown in figure 6, an interface to the AVOPLAN system exists. While the generation of the operation sequence is carried out by the AVOGEN system, time and cost calculations of the determined operations is done within the AVOPLAN system.

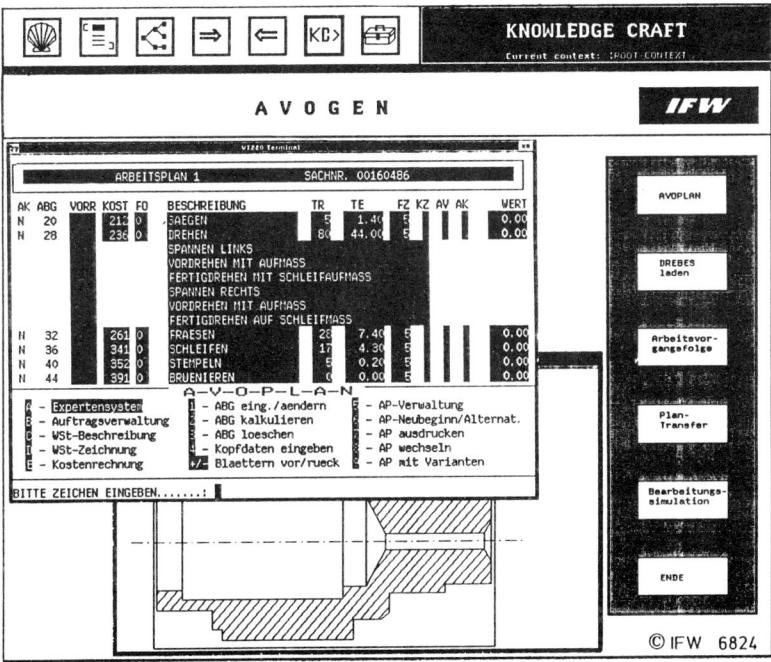

Figure 6: Interface of *AVOGEN*

5. FUTURE TRENDS

Currently, a strong trend can be identified to bring design and process planning closer together on the one hand, and to establish a closer cooperation between process planning and workshop control on the other.

5.1. Integration of CAPP and design

The close connection between CAD and CAPP could be beneficial to both areas. Not only the top-down transfer of feature information generated during the design stage is useful, but also the feedback of information from process planning and NC-tape generation can also greatly increase design productivity [28, 29].

The *SESAME* project (BRITE/EURAM P4539) aims at the development of a simultaneous engineering workstation which will perform design for manufacturing, automatic process planning and NC-program verification in a single integrated system. The major concern is the manufacturing of complex prismatic and cylindrical parts. The project will take advantage of a new hybrid modeller which allows the use of solid models and surface representations concurrently. Having this solid modeller as a kernel, the *SESAME* system will become a multifunctional modelling environment, accessible at any level and allow the integration of the following applications:

- a feature-based design interface with design-for-manufacturing advice,
- a module for mapping between design and manufacturing features and detecting complex interactions,
- a CAPP system, based on AI techniques, coupled to a NC-program generation, simulation and verification system,
- an optimisation package for the final selection of machines, processes, tools, and sequencing of cuts.

In order to achieve the objectives a combination of feature oriented design techniques and complex feature recognition methods will be utilised.

5.2. Integration of CAPP and workshop control

Within the production process, disturbances and capacity bottle-necks often prevent the on time operation of workshop orders, as foreseen in the process plan. In such cases, short-term changes to process plans and corresponding rescheduling have to be carried out [30].

These problems could be eased significantly if it were possible to determine the final course of operations not during process planning, but only during the scheduling of an order in the workshop according to the actual availability of resources.

Firstly, such a concept requires a new structure of process plans. It should provide possibilities to specify, during process planning already, alternative operations and to allow for the technological independence of certain operations ("non-linear process plans").

In order to utilize the increased flexibility of such nonlinear process plans for the compilation of an optimal schedule, the format of process plans has to allow further processing in a computer-aided workshop scheduling system.

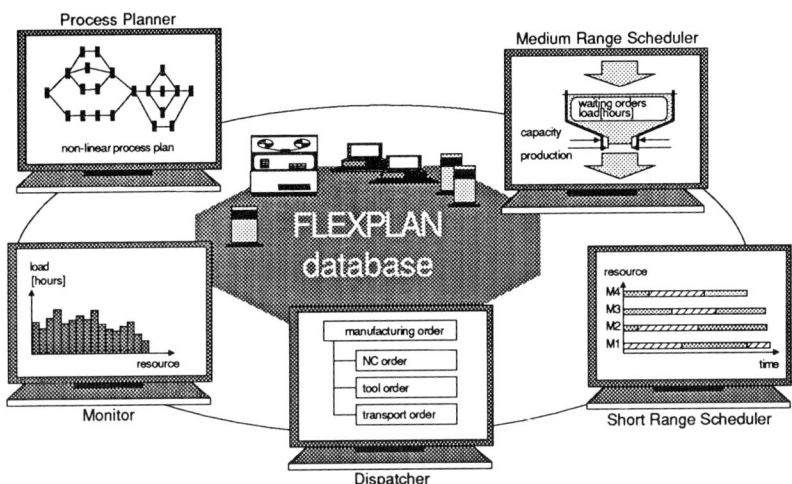

Figure 7. Main modules of the FLEXPLAN System

ESPRIT Project 2457 - *FLEXPLAN* - aims at an integration of process planning and workshop control. The process plan is represented by a net structure and saved in a standard format [31, 34, 7]. Alternative operations are linked by OR-nodes while technologically independent operations are joined by AND-nodes. This results in a net structure, representing all known manufacturing options. The *FLEXPLAN* system will allow for a knowledge based generation of operation sequences on the basis of a feature-oriented product description. For manual process planning the NLPP-editor module has been developed.

The process plan format can be processed by a twofold workshop scheduling module which is also a knowledge based system. It provides functions for both automatic and interactive scheduling and rescheduling. During workshop control the actual situation in the workshop and the state of auxiliary orders (i.e. NC-programming-, tool assembly-, and transport orders) is continuously monitored and fed back to the scheduler. Figure 7 shows main modules of the system.

6. Summary

CAPP research has been directed largely at problems concerning component representations. The use of form features to model discrete 2-D and 3-D components is regarded as a promising tool for the further computerisation of CAD-CAM systems. First attempts tried to standardise methods of describing features (CAM*I, PDES, STEP). Feature catalogues resulted for different workpiece families (rotational, prismatic, and sheet metal parts). Two major approaches concerning the generation of a feature-based workpiece representation emerged. Either the designer has to design with features, or recognition algorithms have to identify features from a conventional CAD representation. Feature-based CAD systems should allow the designer to interactively configure compound features, based on atomic ones.

In the design stage of a product, design features are used to realise functional requirements. During the process planning task the manufacturing requirements are represented by production features, generated from the given workpiece representation. The resulting production feature-based workpiece description is the most appropriate basis for both process planning and NC-programming. It allows the application of a specific planning strategy - the state-space approach - for the generation of the operation sequence.

During plan generation the intermediate workpiece state is analysed to determine which manufacturing activity (represented by an entity called operator) can be carried out next. The application of an operator results in a new intermediate workpiece state. If no operators are applicable a backtracking mechanism is carried out. The planner model selects an operator to be tried next. The total set of operators represents the potential of the real workshop represented by the workshop model.

Recently completed and current projects are presented which highlight results and future trends in the addressed area. The AVOGEN system allows the planning of rotational parts. Within ESPRIT Project 2457 (*FLEXPLAN*) the generation of non-linear process plans is investigated.

ACKNOWLEDGEMENT

The authors would like to thank Klaus Hellberg from the IFW for his relevant and beneficial contribution to this paper.

7. References

1. Anders, N.; Schaele, M.; Prack, K.-W.: AVOGEN- wissensbasierte Generierung von Arbeitsvorgangsfolgen. VDI-Z 132, Nr.4, pp. 49 - 51.
2. Bjorke, O.: Advanced Production System - CAD/CAM for the Future, Tapir Publisher, Trondheim, 1987.
3. Brown, P.F.; Ray, S.R.: Research Issues in Process Planning at the National Bureau of Standards.
4. Cunningham,J.J.; J.R. Dixon: Designing with features: The Origin of Features, Proceedings of the 1988 ASME International Computers in Engineering Conference and Exhibition, Vol.I.
5. Darbyshire, I.;Davies, B.J.: EXCAP - An Expert System's Approach to Recursive Process Planning, 16th CIRP International Seminar on Manufacturing Systems, Tokyo, 1984.
6. Descotte, Y.; LA Tombe, J.C.: GARI: A Problem Solver that Plans how to Machine Parts, 7th International Joint Conference on Artificial Intelligence, Vancouver, 1981.
7. Detand, J.;Kruth, J.-P.;Kempenaers, J.;Pinte, J.;Kreutzfeldt, J.: The Generation of Non-linear process plans. 22nd CIRP International Seminar on Manufacturing Systems, 11.-12. Juni 1990, Twente, Netherlands.
8. Faux, L.D.: Reconciliation of Design and Manufacturing Requirements for Product Description Data Using Functional Primitive Part Features, CAM-I Report, R-86-ANC/GM/PP-01, CAM-I Inc., Arlington, Texas, 1986.
9. Freist, C.: Einsatzmöglichkeiten statistischer Verfahren in CAD/CAM-Systemen, Dr.-Ing. Dissertation, 1984.
10. Granow, R.: Strukturanalyse von Werkstückspektren, Dr.-Ing. Dissertation, 1984.
11. Henderson, M.R.: Extraction of Features from Three Dimensional CAD Data, Ph.D. Thesis, Purdue University, W. Lafayette IN, 1984.
12. Hertzberg, J.: Planen und die Repräsentation der realen Welt. PhD-Thesis, Institut für Informatik, Bonn, 1986.
13. Hummel, K.: An Expert Systems Based Machine Tool Planner for a Distributed Automated Process Planning System, Masters Thesis, University of Kansas, 1985.
14. Husbands, P.; Mill F.G.; Warrington S.W.: Representation, Reasoning and Decision Making in Process Planning with Complex Components, in 'Geometric Reasoning', Clarendon Press, Oxford, pp203-215, 1989.
15. Husbands, P.; Mill, F.G.; Warrington, S.W.: A Knowledge Based Process Planning Systen, Proceedings of the 2nd International Conference on Application of Artificial Intelligence in Engineering, Computational Mechanics Publications, August 1989.
16. Inui, M.; Suzuki, H.; Sata, T.: Process Planning Automation Based on the Manipulation of Form Features, Journal of the Japan Society Precision Engineering, Vol 54, No 10, pp 1903, 1988.

17 Iudica, N.R.: GUMMEX - ein Expertensystem zur Generierung von Arbeitsplänen für die Fertigung, Nachrichten für Dokumentation 36 (1985) Nr 1.
18 Jared, G.E.M.: Recognising and Using Geometric Features, in 'Geometric Reasoning', Clarendon Press, Oxford, pp169 - 188, 1989.
19 Joshi, S.B.; Chang, T.C.: Graph Based Heuristics for Recognition of Machined Features from a 3D Solid Model, Computer Aided Design, Vol 20, No 2, pp 58-65, 1988.
20 Luby, S.C.; J.R. Dixon; M.K. Simmons: Creating and Using a Feature Database, Computers in Mech. Eng., Nov. 1986.
21 Mouleeswaran, C.B.:PROPLAN: A Knowledge Based Expert System for Manufacturing Process Planning. Annals of the CIRP, 33 pp.303-306, 1984.
22 Nau, D.S.: Hierarchical Abstraction for Process Planning, University of Maryland Technical Research Report, SRC-TR 87-105, 1987.
23 Pratt, M.J.: Solid Modeling and the Interface Between Design and Manufacture, IEEE Computer Graphics and Application, pp.52-59, July 1984.
24 Shah, J. A: Scheme for CAD-CAPP Integration. Report, Automation Systems Lab, GE Corporate R&D, Schenectady, NY, August 1986.
25 Shah, J.; M.T. Rodgers: Feature Based Modelling Shell : design and implementation Computer in Engineering, 1988, San Francisco, Cal., USA.
26 Srinivasan, R.; Liu, C.R.: Generative Process Planning using Syntactic Pattern Recognition, CIME, pp 63-66, March 1984.
27 Staley, S.M.; Anderson, D.C.: Using Syntactic Pattern Recognition to Extract Feature Information from a Solid Geometric Data Base, Computers in Mechanical Engineering, pp 61-66, September 1983.
28 Tönshoff, H.K.; Rudolph, F.N.: Neue Ansätze zum Zusammenwirken in der Konstruktion und Fertigung in der flexiblen Konstrukion ZWF-CIM, 84 (5), p.253-257, May 1989.
29 Tönshoff, H.K.; F.N. Rudolph: Wissenbasierte Beratung bei der werkzeuggerechten Konstruktion. VDI-Workshop "Expertensysteme in Entwicklung und Konstruktion", Baden Baden, 13/14 Nov 1989.
30 Tönshoff, H.K.; Beckendorff, U.; Schaele, M.: Some Approaches to represent the Interdependence of Process Planning and Process Control. Proceedings of the 19th CIRP International Seminar on Manufacturing Systems, pp. 257-271, Pennsylvania State University, USA 1987.
31 Tönshoff, H.K.; Beckendorff, U.; Anders, N.: FLEXPLAN - A Concept for intelligent process planning and scheduling. Proceedings of the CIRP International Workshop on Computer Aided Process Planning (CAPP) , pp. 87-106, 21./22. Sept. 1989, published by the Institute for Production Engineering and Machine Tools (IFW) of the University of Hannover.
32 Tönshoff H.K.; Hellberg, K.; Anders, N.; Manufacturing Planning using AI Programming and Planning Techniques. 21st CIRP International Seminar on Manufacturing Systems, pp. 1:95-114, 5.-6. Juni 1989, Stockholm, Schweden.

33 Tönshoff, H.K.;Anders, N.: Survey on State of Development and Trends in CAPP Research within CIRP. Annals of the CIRP, Vol. 39/2/1990, pp. 707-710.

34 Tönshoff H.K.; Beckendorff, U.; Anders, N.; Detand, J.: A Process Description Concept for Process Planning, Scheduling and Job Shop Control. 22nd CIRP International Seminar on Manufacturing Systems, 11.-12. Juni 1990, Twente, Netherlands.

35 Tulkhoff, J.: Lockheed's Genplan. In Proceedings of the 18th Numerical Control Society Annual Meeting and Technical Conference, pp. 417-421, 1981.

36 Van Houten, F.J.A.M; van 't Erve, A. H.: PART, a parallel approach to Computer Aided Process Planning, Proceedings of CAPE 4, Edinburgh, 1988.

37 Woodwark, J.R.: Shape Models in Computer Integrated Manufacture: A Review, IBM UKSC report no. 183, 1988.

GENOA: Feature Based Generation and Optimization of Process Plans

H-J. Held[a] and G. Jüttner[b]

[a]Germany's Application Oriented AI Centre, FAW Ulm, P.O. Box 2060, Helmholtzstr. 16, W-7900 Ulm, Germany

[b]MAHO Aktiengesellschaft, Tirolerstr. 85, W-8962 Pfronten, Germany.

Abstract

In practice existing CAD/CAM-Systems are used to generate optimized NC-programs for rotational symmetric workpieces by using a semi-automized system. For prismatic workpieces the automization of process planning and NC-programming is quite unsolved and therefore one of the most important challenges in the manufacturing industry.

For this reason the gap between Computer Aided Design (CAD)- and NC-Programming-Systems has to be closed by integrated information processes based on a global CAD-/CAPP-/CAM(NC)-data model. The corresponding database has to be used to store the workpiece description by the CAD-System and to retrieve it again by CAPP- and NC-Systems. Additionally, data about existing machines, tools and operations are stored and therefore available for process planning and NC-programming.

These integrated information processes are treated in a common project between FAW Ulm, the Germany's Application Oriented AI-Centre, the tool machine company MAHO AG and the manufactures of automobiles, DAIMLER-BENZ AG and MERCEDES-BENZ AG. Within this project - which is named GENAO - a prototype system has been realized which generates process plans based on a workpiece description stored in a relational database. The workpiece description includes geometrical and technological data which are generated by the CAD-System during the feature-based design process.

Based on this geometrical and technological information a rule based process planning system, named ENGIN, determines tools and operations necessary to machine the workpiece. In addition, ENGIN has an interface to the relational database to access the stored workpiece description.

Additionaly, artificial intelligence techniques are used in an optimization modul in order to optimize the sequence of the determined operations. Thereby the number of tool changes, the number of table turns and the length of tool travel can be minimized.

In summary GENOA is a joint projekt designed to reduce planning time by automization of the planning process and to reduce production costs by minimizing the nonproductive times of the machines.

Keywords

Feature based design; process planning; optimization of operation sequences; artificial intelligence techniques; CAD/CAPP/CAM(NC)-data model; integrated technical information processes; decision table technique; NC-programming.

1. INTRODUCTION

The generation of plans for complex processes is one of the basic tasks in mechanical engineering. This task concerns both the generation of process plans for products to be manufactured internally, and the generation of tenders for customers. When generating a tender, the tool machine company has to determine on which machines and at what costs the customer's workpieces can be manufactured. This determination has to be based on the information given by the customer.

When manufacturing inhouse, the question arises after receiving the internal order. Therefore, process planning is very important not only for technical but also for economic reasons. The planning must be methodically correct and aimed at optimization in order to work out competitive tenders.

The project GENOA (Generation and Optimization of Process Planning) being conducted at the FAW (Germany's Application Oriented AI Centre) in Ulm aims at rationalizing the generation of process plans by means of data processing techniques [5]. An important goal in generating process plans is to optimize the sequence of operations in order to minimize nonproductive time of the machines [3].

Within the project GENOA, which is based on an integrated CAD/CAPP/CAM(NC)-data model, a complete flow of information between the different CA-components will be realized using a relational database. As an example of this process, MAHO AG one of the leading manufacturers of machine tools, has realized automated process planning for a few different workpieces in cooperation with the DAIMLER-BENZ AG and the MERCEDES-BENZ AG.

2. ARCHITECTURE OF THE SYSTEM

The GENOA system consists of the following fundamental modules (fig. 1):

- design
- process planning
- optimization

The design module features (CAD-macros) are used to describe a workpiece functionally and geometrically. All details of information which are relevant for manufacturing (geometrical and technological data) are provided in the design process using pre-determined parameters.

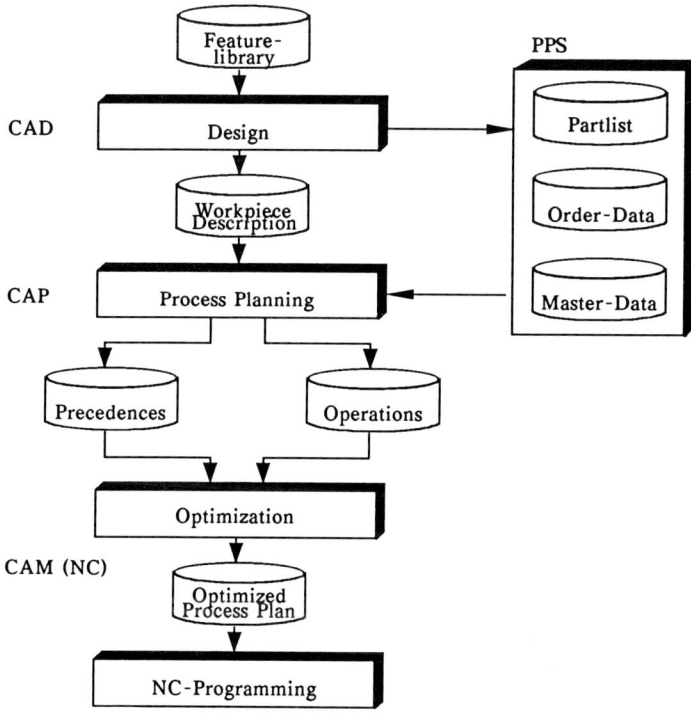

Figure 1. Architecture of GENOA

This description of the workpiece is the input for the process planning module. The data is used to determine the operations for one clamping and subsequently to determine the necessary manufacturing methods, i.e. tools and resources. The process planning knowledge is represented in a rule based decision table system, which is often used in existing process planning systems.

The optimization module has the task of scheduling the operations generated by process planning in an optimized way, regarding not only technological but also economical aspects of manufacturing. The primary criterion within optimization is to minimize the nonproductive times associated with the manufacturing process. These times depend on the number of tool changes, the number of table turns, and the length of tool travel. Within the method database of GENOA, suitable algorithms are provided to determine an optimal operation schedule. This is based on the information about tools and precedences between the different operations gathered by the previously described modules. Unfortunately, conventional algorithms are not able to solve these problems. Therefore algorithms based on artificial intelligence techniques are used in the GENOA method base.

The results of this optimization module can be used as an input for existing CAM(NC)-Systems in order to determine an optimal sequence of NC-operations and thus an optimized NC-program.

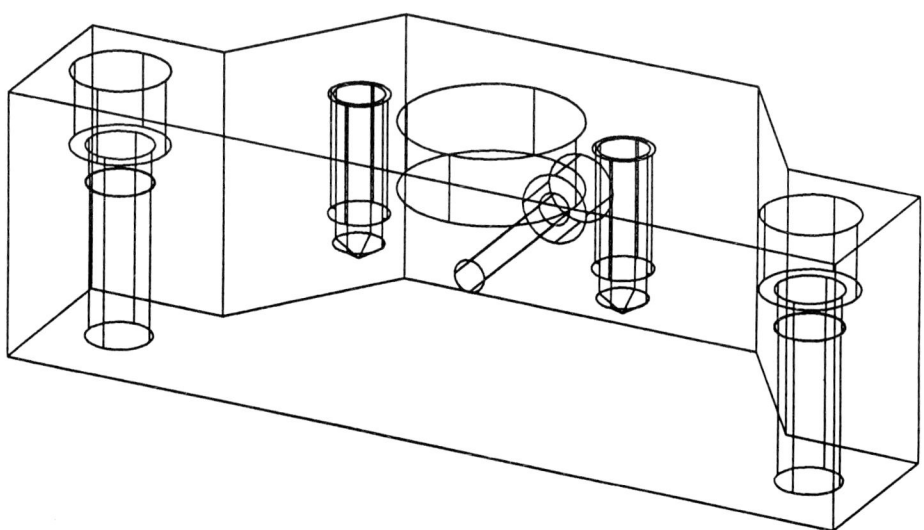

Figure 2. 3-D-model of the workpiece "Bridge"

3. DESIGN

3.1 Feature based design

The realization of an overlapping CAD-/CAPP-/CAM(NC) data model requires a complete geometric and functional description of the workpiece as a result of the design process. The generation of this description is based on the usage of features, the smallest functional units of a workpiece in regard to the manufacturing process (e.g. sink, countersink, thread).

Each feature is described by specific parameters which have significant consequences in process planning. GENOA uses a feature-library which has been integrated in the CAD-System AUTOCAD and thus offered to the engineer. During the design process the user has to select the desired features from the library and then to specify their parameters interactively.

This is further illustrated by the following example:

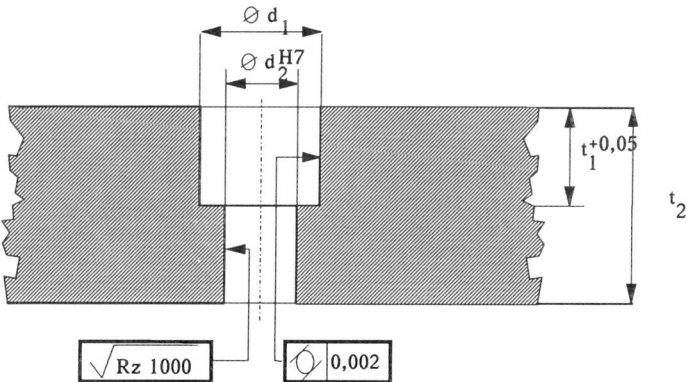

Geometrical parameters
e.g. Diameter d_1, d_2
      Depth t_1, t_2
      Position and orientation data

Technological parameters
e.g. Dimensional toleraces,
      Positional tolerances,
      Surface quality

Processing data
e.g. Name of the feature,
      Type of the feature

Figure 3. Feature "Countersink"

The realized prototype-system includes a feature-library, which has four different feature-types: "channel", "drilling", "countersink" and "thread". An illustration of how the workpiece "bridge" can be described by means of these features is presented in figure 2.

Each feature is described by specific parameters which again have significant consequences in process planning. Figure 3 shows the information provided with the feature "countersink" as an example.

Feature based design is quite different to conventional design techniques. For example, conventional design is characterized by constructing geometric objects i.e. lines or circles with manual insertion of technological parameters (e.g. tolerances). Additionally, CAD-systems usually produce a 2-dimensional (2D) output and are therefore limited in use.

In case of feature based design it is further indispensible for the CAD system to offer a complete functional software environment (e.g. purge, change and output functions). By this way it is also possible to design complex clusters consisting of several features. In case of storing the created cluster in the feature-library the user is able to expand the library and to reuse this cluster in further design processes.

3.2 Determination of Tolerances

During the design process certain specifications (tolerances) have to be determined for each working surface. For this purpose the different features have adequate feature-parameters. The values for these parameters are interactively specified by the engineer. In this context, two kinds of tolerance relations are distinguished: internal and external.

An internal tolerance refers to the accuracy within a boundary surface of feature.

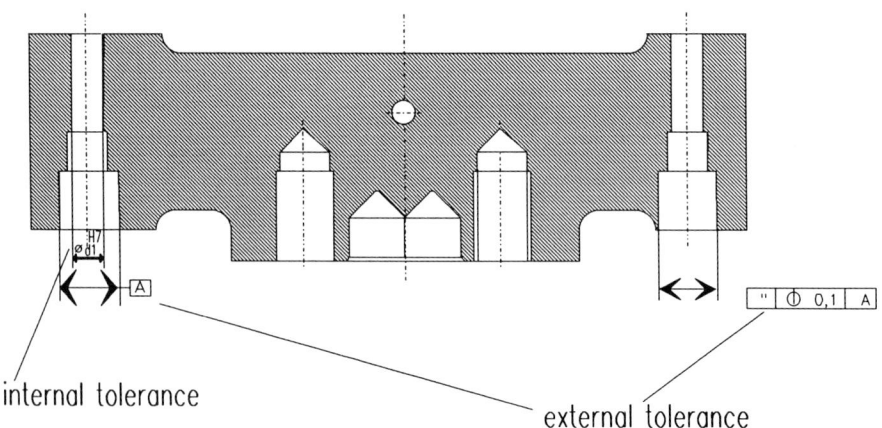

Figure 4. Examples of internal and external tolerances

An external tolerance, however, determine the accuracy between the boundary surfaces of two or more features. This includes, for example, shape and position tolerances.

In order to specify these tolerances, the boundary surfaces of a feature must be individually identified. Figure 4 gives some examples of internal and external tolerances.

4. PROCESS PLANNING

Based on the geometrical and technological description of the workpiece as it is generated during the feature based design process, the process planning module determines the operations for manufacturing the workpiece and also the precedence restrictions, which must be considered during manufacturing. A rule based knowledge representation technique is suitable for both tasks, therefore a decision table system which is often used in practice is applied within this module [4].

4.1 Determination of Operations

The process planning module requires the following input-data, which are available in the CAD/CAPP-database:
- Description of workpiece (consisting of features and feature-parameters)
- Master data (e.g. tool data, machine-specific parameters).

Using workpiece descriptions and the resource master data, the necessary machining operations and tools are determined. As essential requirement for process planning is the availability of technological information in addition to the geometric information. By applying different types of rules this technical information influences the determination of necessary operations. This can be shown by using tolerances as an example. Thus, a drilling with the diameter of 40 mm and a depth of 50 mm with free size tolerance (DIN 7168) results in the following operations:

- (1) drill
- (2) countersink.

However, if the accuracy "H7" is specified for the same drilling, additional operations must be inserted:

- (1) centerdrill
- (2) drill
- (3) boring
- (4) reaming
- (5) countersink

Consequently, a result of the process planning module comprises the set of all operations which are necessary to manufacture the workpiece from a given unmachined part.

The operation sequence of the generated process is random ordered and therefore not optimized. The optimization module (described in section 4) is able to determine an optimal sequence, if the restrictions (precedences) that result from the manufacturing technology are available.

4.2 Generation of Precedences

In general, a precedence between the operations A and B expresses that operation A must be finished before operation B can start. These precedences are influenced by geometrical, technological and company-specific manufacturing restrictions. For instance, there has to be a centerdrill operation before drilling a hole or there has to be a boring operation before reaming in case of accurracy "H7" (see above example). Usually, these operations will not be carried out directly one after another. In order to minimize the number of tool changes many operations using the same tool may be executed between operations A and B.

The knowledge for the determination of the necessary precedences is also represented by rules within the above mentioned decision table system.

5. OPTIMIZATION

The task of the optimization module is to find an optimal sequence of operations in order to minimize the nonproductive machine times. The different operations and precedence restrictions determined by the process planning module are used as input to this module. The aim of the optimization process is to minimize nonproductive times related to manufacturing and thereby to reduce the number of tool changes, the number of table turns, and the length of tool travel [1].

Unfortunately, there is no known conventional algorithm to calculate the optimization problem which solves this problem in polynominal and therefore acceptable time, because these algorithms are based on calculating exact solutions. On the other hand there are well known methods based on artificial intelligence technique which are able to solve these problems by using heuristics and estimations. Consequently these algorithms don't guarantee an exact solution. The quality and runtime of these optimization methods depend on the problem. In practice, however, the engineer is often satisfied with an approximate solution which is calculated in few minutes. In some other cases he will need a more exact solution and therefore he'll probably accept a runtime of a few hours.

In general, users are not able to choose a suitable algorithm that efficiently solves their problem as exact as desired. For this reason GENOA uses a socalled method database in order to intelligently and automatically choose an appropriate method based on the user input and an analysis of the problem.

The input information for this method database consists of the user's optimization problem (for example classified by the number of operations or the number of precedences) and the specific requirements for the optimizing module. By utilising this method the user can choose between obtaining an efficient solution and a short time or can enhanced and better solution which requires a longer system-runtime.

An optimized process plan is characterized by the fact that operations with the same tools are usually carried out one after another even if it is necessary to turn the table.

However, if for a special flexible manufacturing system the time for changing a tool is shorter than the time for turning the table, it may be better to first carry out all operations on one surface than to turn the workpiece. This implies that it is necessary to change the tool more often. If this information is evaluated by existing NC-programming systems, a better automization of the CAD/CAM-coupling will result and, moreover, an optimized NC-Code will be generated regarding the sequence of operations.

6. DATA MODEL

6.1 Characteristics

The integration of the above described modules is carried out with the help of a distributed relational database. To satisfy all requirements of the different applications, concurrent write accesses to the corresponding database are necessary. Therefore highly sophisticated access methods and transaction mechanisms are necessary to guarantee consistency and integrity constraints of the database. For this purpose the use of a two phase commit protocol will be applied within the GENOA project [2].

The distributed CAD/CAPP/CAM-database is characterized by a common conceptual data model. The entity-relation model (ER-model) is suitable for this purpose due to its simple and clear architecture. The ER-model consists basically of entity- and relationship-types. All entities can be subdivided in additional subtypes.

In the following two chapters, the most important entity and relationship-types for the developped CIM data model is being described. In figure 5, an overview of CAD/CAPP Data Model is given.

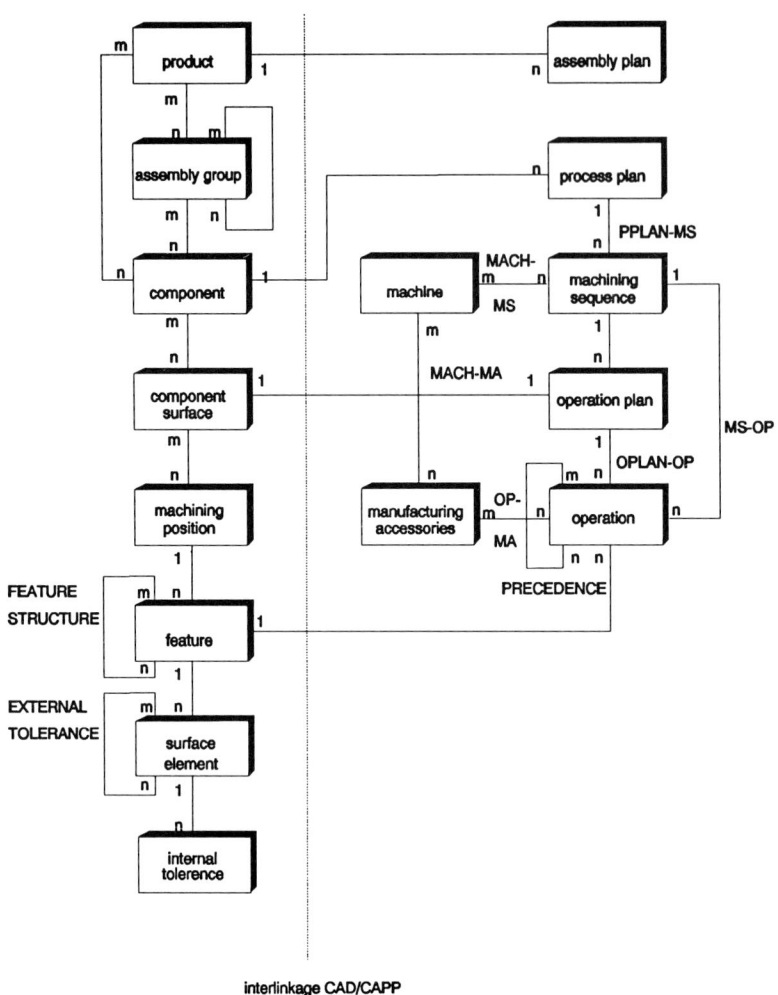

figure 5. CAD/CAPP Data Model

6.2 CAD Data Model

The data used in the design module are represented in the following data model:

During final assembly, a PRODUCT is being mounted together out of several individual COMPONENTS and ASSEMBLY GROUPS. An individual component represents a workpiece which is machined on a MAHO machine tool. For this reason, this component is described as workpiece within the CAD database. Each component is limited by different COMPONENT SURFACES. On a component surface there are several MACHINING POSITIONS such as stepped holes, counterbase, pockets, slots, stepped surfaces etc.

A MACHINING POSITION is being designed with the aid of one or several features. These features are used by the engineer as FEATURES, which are very important within the CAD/CAPP-process.

The hierarchical structure of these features is represented by different layers of the CAD-system. Thereby the features corresponding to the different layers influence the determination of operations necessary to process complex machining positions. One example would be a stepped surface with a stepped hole, whereby several features are added together. In every feature represented in the database one or several adjacent SURFACE ELEMENTS are assigned. A typical example is the cylinder and bottom surface of a hole. The surface elements contain several parameters like surface details and tolerances, which are stored in the corresponding entity-type INTERNAL TOLERANCE and relationship-type EXTERNAL TOLERANCE.

6.3 CAPP Data Model

The data used in the modules process planning and optimization is respresented in the following data model:

For the manufacturing of a workpiece a PROCESS PLAN is needed. This plan contains operations of all machining positions including all component surfaces, which have to be executed in all clamping arrangements on one or several machine tools. Depending on the batch size, several process plans may be neccessary.

One process plan again consists of several MACHINING SEQUENCES. One machining sequence contains all operations of every machining position of every component surface at one clamping arrangement at one machine.

One machining sequence consists of several OPERATION PLANS. One operation plan contains all operations of all machining positions which are neccessary to handle one component surface at one clamping arrangement at one machine.

One operation plan again consists of several operations whereby an operation can be considderd as an elementary partial machining of a machining position with one cutting tool.

Within the database a devision between the design and the manufacturing process is indicated by the interlinkage between CAD and CAPP relations (see dashed line in fig. 5).

7. CONCLUSION AND FUTURE WORK

Most available systems for process planning do not offer a direct interface to a CAD system. Thus, it is necessary to input the workpiece description manually, which is both rather time cosuming and more susceptible to errors. Other systems do not offer the opportunity to optimize a process plan with regard to specifiable criteria. GENOA is a system that overcomes both limitations.

The integration of the modules within the whole system is realized using a relational database. This database stores the data generated by the different modules (design process, process planning, optimization) and also makes the data available to the other modules. By utilising this process an integrated flow of information can be achieved.

Within the next phase of the project, the main research goals are the improvement of the interface to CAD systems and the realization of an interface from the optimized process plan to NC-programming systems. By this process, a complete flow of information from the design stage via process planning through optimization and NC-programming is achieved. Beyond this the qualitiy of the generated NC-programs is improved.

GENOA uses the CAD-system AutoCAD within the design module. For process planning the decision table system ENGIN is used, which runs on an Apollo workstation under the UNIX operating system. The same workstation is used for optimizing the operation sequences. Methods and method base used by the GENOA prototype-system have been implemented in the programming language C. The user interface of the system is based on X-WINDOWS and OSF/MOTIF. The integrated flow of information has been realized using the relational database system INGRES.

8. REFERENCES

[1] Feller, H.; Held, H.-J.; Jüttner, G.: Mit KI-Methoden Arbeitsgangfolgen planen und optimieren. In: Arbeitsvorbereitung AV, 27. Jahrgang, 2/90, March-April 1990, p. 59-61, Carl Hanser Verlag, München, 1990

[2] Güntzer, U; Held, H.-J.; Jüttner, G.: Verteilte Datenverwaltung mit INGRES, TRANSBASE oder ORACLE. Konzeption einer CIM-Datenbank am FAW. FAW-report No. B-90005, Ulm, 1990.

[3] Held, H.-J.; Jüttner, G.; Feller, H.: Generation and Optimization of Process Plans. In: (Österreichischer Ingenieur- und Architekten-Verein, ED.), ISATA, 23rd International Symposium on Automotive Technology and Automation, Vol. 2, Austria, 3.-7. December, 1990, pp. 249-261, Vienna, 1990.

[4] Hüllenkremer, M.: Rechnerunterstützte Arbeitsplan-Erstellung in einem Entscheidungstabellensystem, In: (Techno Congress GmbH, ED.), Neue Wege der rechnergestützten Arbeitsplanerstellung. Techno Congress, Karlsruhe, March 1991.

[5] Jüttner, G.; Feller, H.: Entscheidungstabellen und wissensbasierte Systeme. Anwendungen in der Arbeitsplanung. Eine Studie der Nixdorf Computer AG am FAW Ulm. Oldenbourg Verlag, München, 1989.

Planning of Operation-sequences With an AI-based Knowledge-acquisition Tool [1]

Hans G Vogt[a] and Per Zaring[b]

[a]Department of Production Engineering, Chalmers University of Technology, Gothenburg, Sweden

[b]Department of Information Processing, Chalmers University of Technology, Gothenburg, Sweden

Abstract

The paper describes the theoretical background and an example of practical realization of an operation-planning system for manufacturing. It defines the cutting depth as the key-variable in cutting operations and outlines the way in which knowledge-acquisition by operator-input can be transformed in production-adapted normalized data. The main strategy for planning shows to be backward-chaining using customized relational databases combined with expertsystem-facilities. Flexibility and reliability is enhanced through minimization of the information content by using tolerated values.

1 Introduction

In the field of producing parts there are three distinct stages as steps towards realization:
design,
planning and
manufacturing.

In the following the principals of design will be explained some further because they show the matter of planning in general. The fundamentals are identical with principals for planning of manufacturing processes which will be dealt with, though in the latter case the methods of problem solving are more difficult to apply.

---

[1] This work is supported by the Swedish National Board for Technical Development

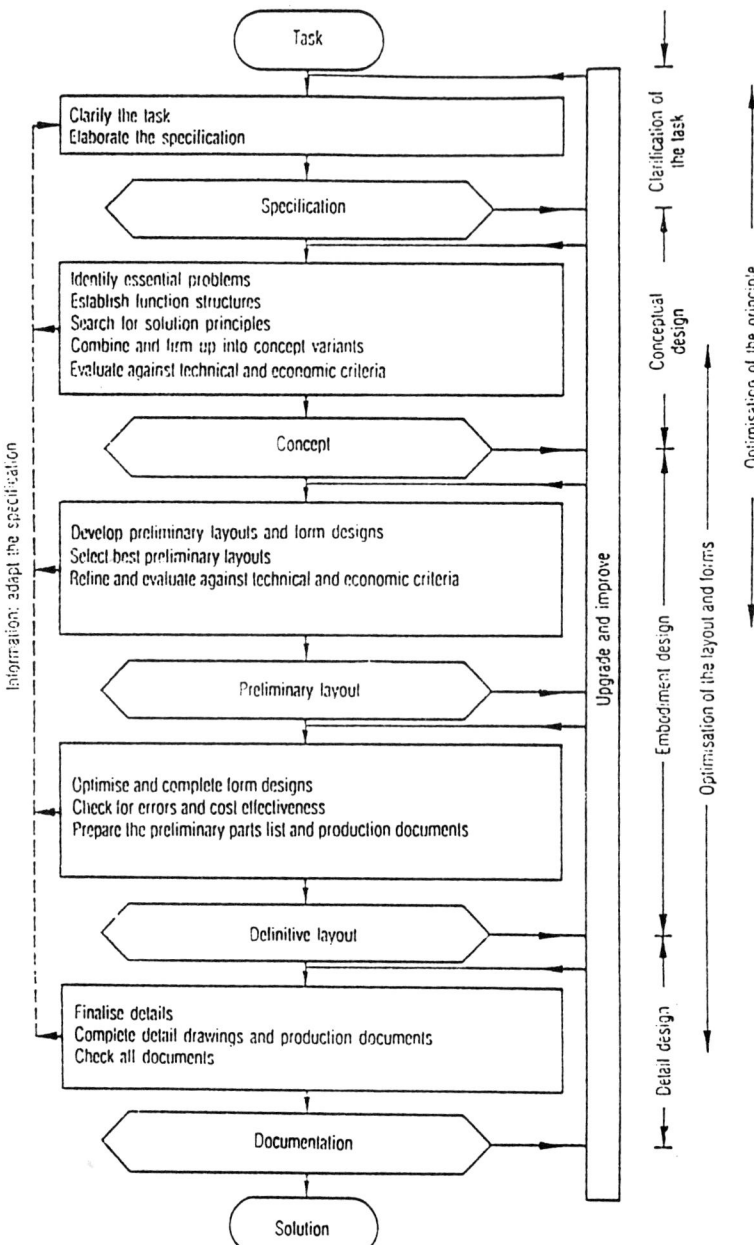

Figure 1   The design process (1)

The purpose of this paper is also to give details about the way to structure knowledge for the design and the planning process in CAD/CAM-systems which should provide the designer with manufacturing knowledge.

As an example the planning of holemaking operations with a newly further developed expertsystem will be described in the AI-perspective of capturing knowledge. The manufacturing of the feature "hole" with size tolerance, surface roughness, positional tolerance etc asks for complex machining usually involving several steps of operations. This claims for expert knowledge.

1.1 Design, planning and manufacturing

In the design phase the demands for form (geometry) and properties of the part are determined. This is done in a design process in several steps each containing specific knowledge. See the Pahl & Beitz model (1) in figure 1. These activities are connected to customer demands as shown in figure 2, but they have to take in account the possibilities of the production resources to realize the product. Figure 3 shows these dependencies. This is the reason why the usage of these resources is determined in the design stage at 70-90 % with the planning stage as the link.

Initially we can distinguish 3 types of design each of them reflected in later steps in the same way:

original design,
adaptive design and
variant design.

Original design is elaboration of an original solution principle for a product or system with any type of task. *Adaptive design* involves the adaption of a known system with the same principal solution to a changed (similar) task. This is often accomplished by recall of the original design. *Variant design* involves the variation of size or arrangement of components in a known solution. The principal solution is the same.

1.2 General methods and working matter

At each level of the planning system two methods are used in the same way. The *select method* is used to select the actual area of interest either a working level, a model or a specific object. The *create method* either helps to manually create new manufacturing processes to our solution of the actual requirement or automatically searches for solutions and gives a number of alternatives. In the case of using the create method in a manually way the *compose method* is used which arranges the objects in to a solution model. In the case of automatic generation of process plans there is a need for a method to search the best alternative.

The application of these methods will be explained in terms of problem-solving methods in chapters 1.4 and 2.

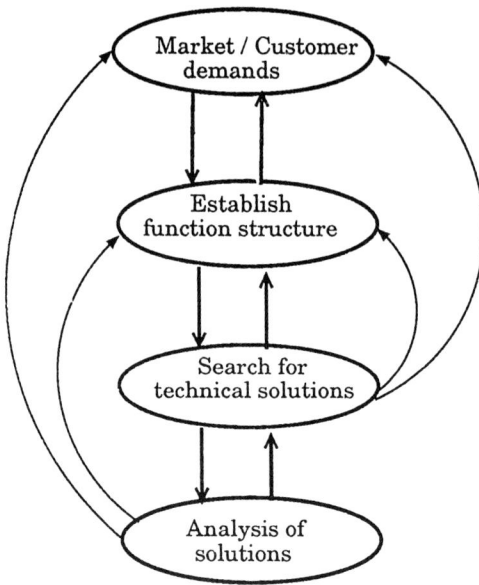

Figure 2   Dependencies and methodology of design (Pahl & Beitz)

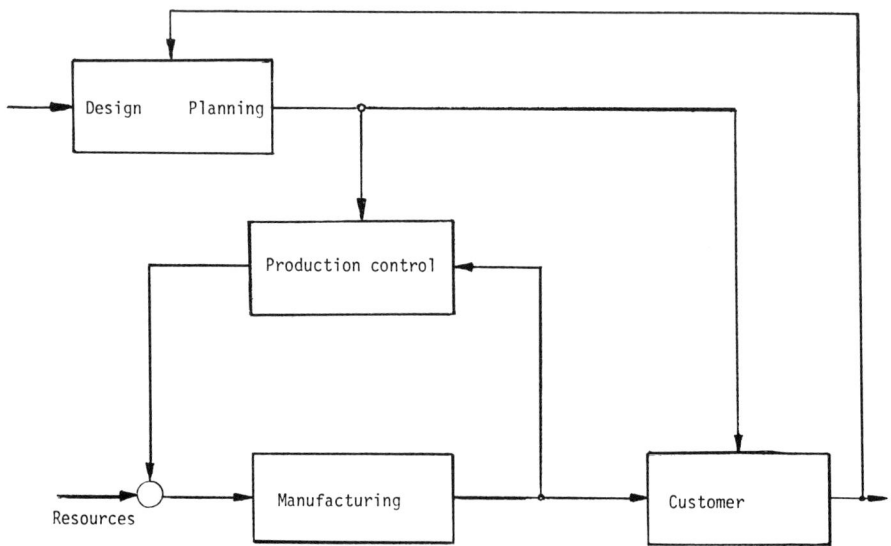

Figure 3   Adaption of manufacturing processes

In general two different working matters can be applied:
heuristic rules and
axiomatics.

The use of *heuristic rules* for selecting the most suitable solution is not very successful when generating new process plans where a high degree of creative work is needed (generic planning). On the other hand the adaptation of known solutions to a new problem (adaptiv planning) is much easier when using heuristic rules. They provide a solution, not necessarily the best but acceptable as proven. That means, that they are based on experience which usually has been collected under a considerable amount of time. The disadvantage is that they restrict the planner to think of previous solutions instead of trying to find new principles to solve the problem.

The *axiomatic approach* gives a general method or forms the outlines for such a method specially appliable to generic planning. Some general principles have been formulated in connection to the design process by Suh (2):

Axiom 1 (Axiom of Independence): Maintain the independence of functional requirements.

Axiom 2 (Axiom of Information): Minimize the information content in a solution.

The independence axiom postulates that it is preferable to have a separate planning solution for each functional requirement. It is also preferable to have each solution of a planning step independent of other selected planning solutions in the system.

The information axiom states that the solution (process plan) with the minimum amount of information should be chosen when several solutions are possible. The information content can be measured as
$C_i = \log_2$ (range / tolerance).
Range is the range on planning solutions and tolerance is the tolerance on functional requirements.

At this time it is not decided wether the design process has to be performed only by the designer or with help of a computer system. The use of the computer has the precondition of exact definition of the singular steps in the process. This is valid also for the activities in the manufacturing phase. In the following the use of computer aid is assumed.

1.3 Implications of taxonomy

A computer system handles symbols and has no understanding of the meaning of these symbols. Therefore it is of great help in using these symbols by applying a pattern to them i e using taxonomies. By classifying concepts into taxonomies a concept can at least be unambiguously interpreted in relation to other concepts. That gives a semantical meaning to a symbolic description of a concept. In a system for design one of the major knowledge sources will be the developed taxonomy of engineering concepts.

In the view of designing planning systems an important task is to design taxonomies similar to these concepts i e design taxonomies applied to another domain. That may result in a representation of the same concept in different taxonomies and the avoidance of misinterpretation (inconsistancy) in the design/planning system.

A reflection of this is the classification of a concept from different perspectives, so that a concept is given by the specifications in all perspectives. In figure 4 features are classified in a design perspective and a manufacturing perspective. Both of the classifications provide a concept where equal concepts can be identified by bridges (3). The contents of a concept is then derived from both classifications through the bridges.

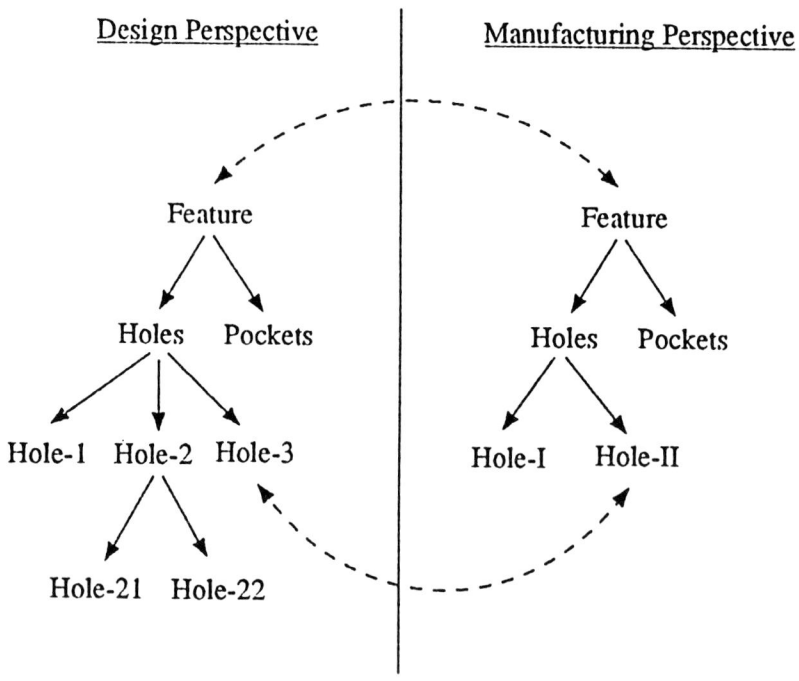

Figure 4   Concepts in multiple perspectives (3)

In the later presentation this is accomplished by using the design specifications of a hole to the planning of operation sequences for manufacturing. At this stage of development these are the following:
> **nominal diameter,**
> **tolerance grade,**
> **tolerance position,**
> **positional tolerance,**
> **surface roughness** and
> **type of material.**

In the computer system (CAD/CAM-system) the bridges are programs extracting the feature definitions from the design concept and providing them as input to the planning system. They are the goal variables in the closed circuit from design, planning, manufacturing to the customer. Building and maintaining the unified taxonomy of desing and manufacturing is not an easy task as the efforts of CIRP (4) and others show.

1.4 Problem solving

In general there are two different methods to use rule-based reasoning for planning i e searching steps to reach a goal state:
> **backward chaining** and
> **forward chaining**.

The process of *backward chaining* is the attempt to solve a problem by stating a goal and looking in the database for the conditions that would cause it to come about. Then the process iterates by using those conditions as the goal, searching for their preconditions, and so on. In other words, the inference engine endeavors to find a value for an overall goal by recursively finding values for subgoals. This method is also called "**goal-related search**".

Reasoning with *forward chaining* is the type of activity done in a system that applies operators to a current state in order to produce a new state, and so on until the solution, the goal is reached. In other words, the problem-solving technique is characterized by working forward from known facts towards conclusions. This method is also called "**datadriven search**".

2 Start and goal for planning

The aim of process planning is to develop a plan of the manufacturing process. The result is a complete sequence of machining operations each of them creating a part-feature closer to the goal state. In the manufacturing world this is done by using a tool.

In the intermediate state, however, the product has to be created from raw material. So, the **start state** of the planning process is the concept of a **blank**.

In the case of mechanical engineering the geometry (feature) of the tool is superposed to the raw part by intersection. Tool geometry and tool movement create a 3-dimensional body which describes a lower limit for the shape of the raw part. The upper limit is given by the maximum load which the tool/machine can handle i e the maximum allowance of the intersection. In daily speech this is called the **cutting depth**.

The intersection is the material to be removed, in the case of cutting operations this is the chip volume. So, if the start state minus the load is less or equal the goal state then the goal state can be reached by one single operation.

Now, the **first question** is: what reasoning method does the planner use when the planning of operations and operational-sequences has to be done?

> In the first place the planner will use **rule-based reasoning**. This is the only possibility to deal with the vast amount of possible solutions. He will compare goals with conditions.

The **second question** is: what type of rule-based reasoning should be used to solve the planning problem?

> In the second place the planner will use **backward chaining**. The design specifications define the goal state and he will look for a manufacturing process as such (the conditions) which can produce a part of this specific kind.

Backward chaining as the type of planning is in coincidence to the **axiom of information**. The design specifications represent the minimum of information for the planning process. Design for assembly (DFA) by Boothroyd is an example for the successful application of backward chaining.

Forward chaining is excluded, because the planner will never be able to scan all the possibilities the production processes provide to make a part starting with the blank, the raw part. An investigation performed at the department showed at least 460 different methods of manufacturing, many omitted a o the groups of methods for assembly, welding, soldering, painting, hardening and so on.

However, in respect to the start state with a blank, the planner will switch in the **intermediate stage** to **forward chaining** evaluating the load i e intersection which the chosen method or tool can provide. He will then decide if the applicable range of load allows him to reach the goal state from the start state, the blank. If not, he establishes a new goal state using the allowance backwards to come nearer to the start state.

Other aspects of problem-solving concerning the type of knowledge in manufacturing in general and hole-making operations in special have been discussed in a previous paper (5). One of the major experiences is, too, that the planner developes working matters/algoritms by transforming rules into procedures based on experience.

3 Working example

3.1 Principal concept

A computer based decision support system for planning of hole-making operations has been developed and presented in previous papers (5), (6). It is called IHOPE, Interactive Hole Operation Planner Expert. The strength of the system emanates from the robust design of the functionality giving tolerated values and on the interactivity which provides extensive knowledge-acquisition capabilities. This is achieved by using different update-mechanisms during the building of the initial knowledge-base and during the planning sequence. A brief description of the system will explain it.

In short, the knowledge base is a set of relational databases for data on the performance of tools/manufacturing methods, for data on operation sequences and a set of rule-bases for dealing with certain issues. So, we have the two major data-manipulation strategies present. The knowledge base is manipulated by performing procedures and by invoking the inference engine.

It should be mentioned that system works mainly procedurally but uses many ordinary if-then-statements representing transformed expert knowledge.

3.2 The planning session

Initially the planner is asked if he wants to make a plan for manufacturing a hole or to update the rawdatabase or the reportdatabase. We presume that the concept of a hole is present.

In a form the planner is asked to enter the following data:
tolerance grade: If the tolerance grade is not known then it is asked for the size deviation expressed as upper and lower limit size. The system then calculates the tolerance grade. If these values are not known either then the system establishes the tolerance grade IT13, known as "factory tolerance".

surface roughness

nominal diameter

material group: If the material group is not known then the planner can proceed to the next step, the input of steel standard number. The system then looks for the appropriate material group and inserts it.

steel standard number: see above.

The system now scans the reportdatabase for previous solutions i e operational sequences. By using diameter ranges instead of exact values the RDS can retrieve similar plans, perform **adaptive planning**. The diameter ranges are arranged according to the ISO-tolerance system.
The planner can accept some solution or proceed to a create a new solution. Let us presume that he will do the latter..

Some preliminary questions follow dealing a o with **tolerance position** (the lower size limit). This will be used to correct the nominal cutting depth for the actual value of the operation. This is done by using a rule-base. If the tolerance position is not given then it is set to "H" i e the lower size limit is 0.

Input of the **positional tolerance** initiates a rule-base which restricts the domain of possible solutions for instance to tools declared as "stable".
Other additional values can be inserted dealing with the type of machine-tool, speed etc, each restricting the search domain.

At this point the **decision** has to be made which tool/method to use to fulfill the functional specifications. This is done by selecting the tools/methods having the performance according to the above defined goal values.

Performance data for tools/methods based on **experience** are stored in the **database**. This is described in (5) and chapter 3.3. See figure 5 for manual use, described in previous papers, too. The information content is minimized in several ways:
    dividing the total diameter range into categories and
    setting tolerances for nominal diameter
on basis of experience structured by the International Standard Organization (ISO) System.

The system *selects matches* and *presents* the tools as *candidates* to the planner. This is done in a special way taking in account the *conditions* for reaching the goal as follows.

The functional requirements of a tool for machining are expressed by the load, the cutting depth. Of all possible values there is a range of practical solutions for each tool. There are a lower limit (minimum cutting depth) and a normalized value (nominal cutting depth) for that. This issue has been discussed in detail in previous work (7).

On the basis of these values a normalized value of the *prefinishing tolerance* can be calculated. The values are stored in the database, too. They represent a tolerance of the start state of the operation. Once again, by using a tolerance of the cutting-depth (the cutting-depth variation) the information content is minimized.

Consequently the candidates for the (finishing) operation are presented in the order of their cutting-depth-variation or prefinishing tolerance respectively.

Now, the planner can *accept* one of the tools on basis of the *default data*. He checks the blank specifications with the start state definitions for instance by calculating the prefinishing diameter and comparing it with the diameter of the blank. The system will provide this function, too. If a fulldrilling tool is accepted then the planning always is done.
If the start state is not reached then a second search, the search for an *prefinishing tool* is initiated. That search is performed in exactly the same way as for the finishing tool..

On the other hand, the planner can *accept* the tool as reaching the goal state but from a *different start state*. In that case he can initiate a calculation of the start state by **giving own values** for the cutting-depth variation or the cutting depths.
In this case the **tool** is marked as **modified** and can be stored separately in the **rawdatabase**. Another option is that this procedure can be repeated.

The further steps are quite obvious. Tools/methods are combined to a **operation sequence** until the start state of the blank is reached. Then this sequence can be stored in the report database.

3.3 Decision support for economical considerations

The system provides a decision support for economical considerations to the planner, too. For that purpose the tools/methods are characterized by the
relative cost value.
At this state of development relative cost values are given as function of the nominal diameter and are stored in the rawdatabase.

The relative cost values are evaluated by using a kind of "calculation of equivalent" (parametric calculation method) by using the actual cost of a certain machining operations as the base value.

During the planning session the planner is asked for lacking values and can assert them to the database. The system will assist him in the necessary calculations.

Thus, the operational sequence shows technical data as well as economical data. As separate functionality the tools can be displayed in order of their economical value.

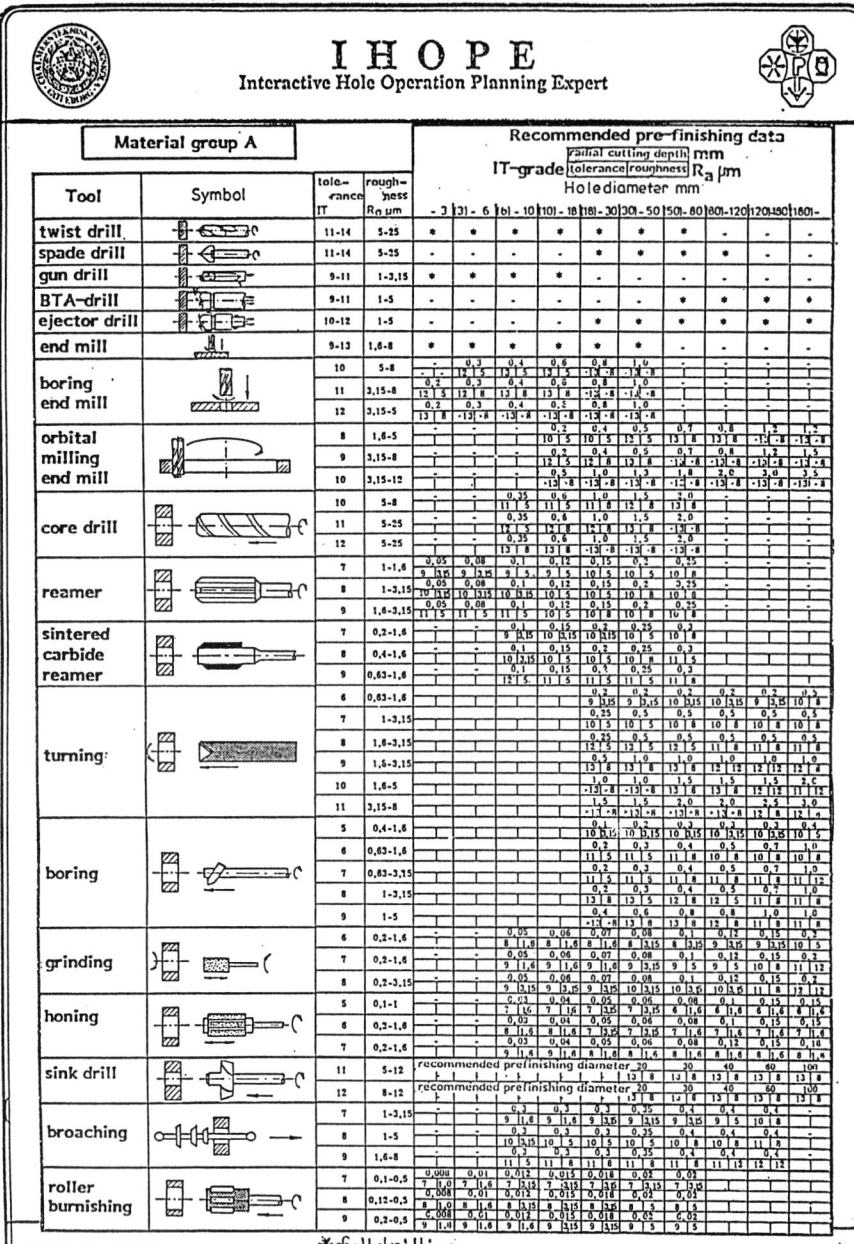

Figure 5 Decision sheet for manual planning of holemaking-operations

3.4 The update session

The update of the databases can be described in short as consisting of the following items:
inserting, deleting, editing of
   base data (goal state data, finishing data, performance data),
   conditions (start state data, prefinishing data) and
   sequences (planning solutions).

In this way a o new tools/methods can be added to the database even during a session as described above. The insertion of operation sequences shall be explained in more detail.

The usual way to present operation sequences is in order of proceeding. They appear as datadriven. This is familiar to the operators at the shop floor, so, the update system for insertion of sequences is designed in the same way.

The **operator inserts** the tools/methods in order of appearance by picking them from a menue of available tools/methods. In the next step the operator is asked for and describes the nominal and actual *values* for instance cutting depths for the *first tool*. The *system calculates* the *goal values* and provides them as the *start values* for the *second tool*. It can calculate the start values for the blank, too. The operator can compare these start values with his own data and thus *controls* the correctness of the previous inputs.
If this is the case then the operator/planner will proceed and the procedure continues to the second tool, inputs are repeated and so on until the complete sequence is inserted. Figure 6 and Figure 7 show prints from an input-session with IHOPE.

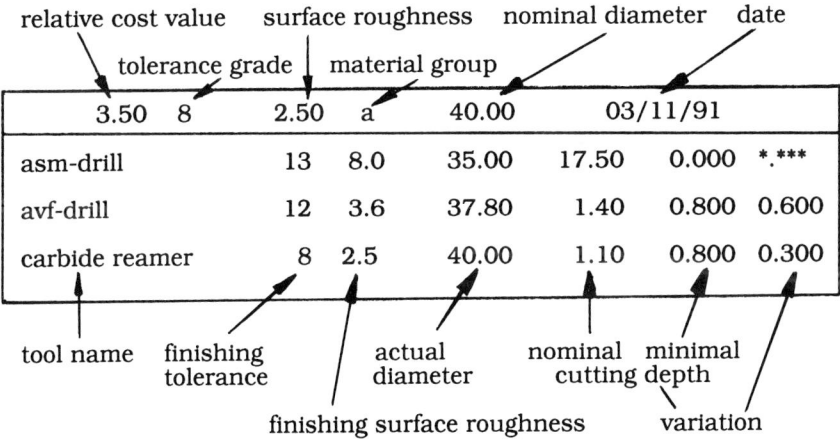

Figure 6   Report of an operation sequence for holemaking with IHOPE

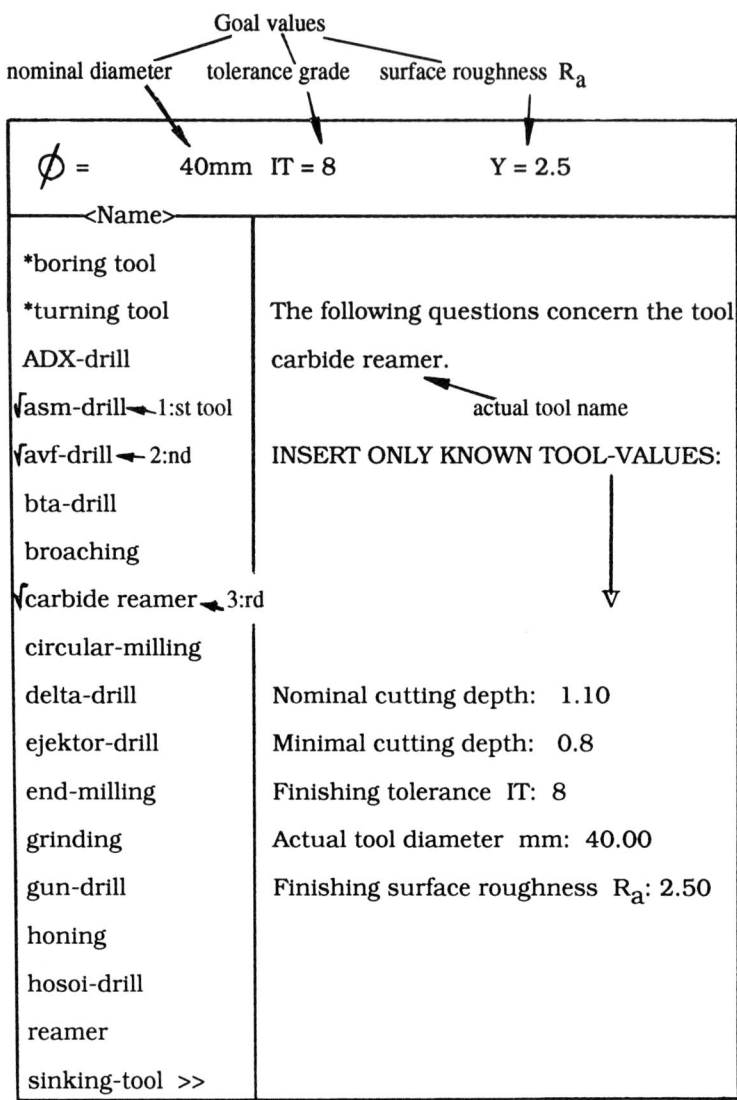

Figure 7 Input form for tools i e a carbide reamer, 3:rd tool of the operation sequence with IHOPE in figure 6

4 Conclusions

A decision support system for the design phase dealing with the producability should be based on experience. The problem-solving method should be goal-driven. The planning cycle should include interactive knowledge-acquisition capabilities. Databases should be based on parted domains and tolerances which should be given separately for each functionality. The use of evaluation functions in the database system should permit to compress and extract informations. After a certain time a planning system based on these principles can become autonomic i e perform automated planning based on minimum information content.

A working example of a planning system for the feature "hole" is presented which carries out the above mentioned tasks. It has been developed by using an expertsystemshell combined with a relational database, spreadsheet and procedural environment.

5 Bibliography

(1) Pahl, G. & W. Beitz: Engineering Design, a Systematic Approach.
Springer Verlag, Berlin, 1984.
(2) Suh, N.P.: The principles of design.
Oxford University Press, New York, 1988.
(3) Marino, O., Rechenmann, F. & P. Uvietta: Multiple Perspectives and Classification Mechanism in Object-oriented Representation.
Proceedings ECAI 90, Stockholm, 1990.
(4) CIRP: Unified Terminology. Part I to Part V.
CIRP, Rue Mansart, Paris, 1986.
(5) Vogt, H.G. & P. Zaring: A Computerized Interactive Hole Operation Planner Expert (IHOPE).
Proc Expert System 90, British Comp Soc, Addis/Muir, Cambridge Press, ISBN 0-521-40403-7, 1990.
(6) Vogt, H.G.: Expertensystem plant Arbeitsgangfolgen:
AV 27 (1990) p 134-136.
(7) Vogt, H.G.: Beeinflussung der Zerspanung bei der Innenbearbeitung insbesondere beim Gewindebohren.
Diss, Chalmers Univ of Techn, Gothenburg, 1985.

OPERATIONAL PLANNING FOR COMPLEX MACHINING

CHAIRMAN: M. SZAFARCIK
POLAND

NC-Programming Systems for Multi-tool Cutting Lathes

A. Storr and Th. Reibetanz

Institut für Steuerungstechnik, Universität Stuttgart, Seidenstraße 36, 7000 Stuttgart 1, Germany

1. Introduction

In the last years the change-over from series to order production leads to the development of new machine tool concepts like CNC-multi-tool and CNC-multi-spindle cutting lathes /1,2/. The major aim of multi-tool cutting lathes is the complete machining besides the possibility to carry out e.g. drilling and milling machining, too. They meet requirements concerning short machining, preparing, lying and thus production and delivery times (fig. 1.1). These advantages are often opposed by not existing programming functions in order to benefit from the machining possibilities regarding time-optimal production.

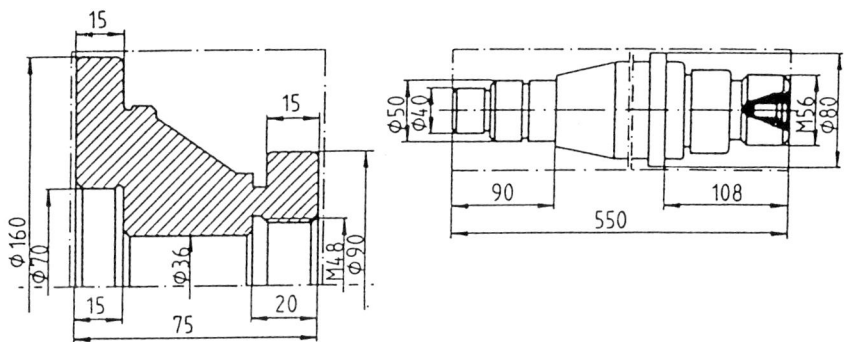

Figure 1.1. Examples for Saving of Piece Rate with Multitool Treatments /3/

In order to avoid these disadvantages, a system concept is presented which, tailor-made for the requirements of new cutting lathe concepts, allows an optimized procedure in the computer-supported NC-programming. It stands out for a division into different levels which contain specific user-directed functionalities which are called system components.
The user may dispose over functions for the part programme processing, for the NC-data processing as well as for the treatment simulation. These system components represent functions which can be automated with the help of conventional calculation methods. The preparation of the workpiece data and mainly the part programming is done with the help of a computer but little automized since here a lot of heuristic knowledge has to be processed in a ruleorientated way. The use of form features with the help of knowledge-based methods is proposed as a solution for the further automation of the workpiece preparation as well as of the part programming. Figure 1.2 compares analogous expressions of conventional application programming- and expert systems.

	Application programming system	Expert system
Programmed in:	arith./log. programming language	symbolic, object-oriented, arith./log. programming language
Formulation of the problem solution	application programming language	types of knowledge representation (rules, frames, objects)
Support of the system generation	editor	knowledge aquisition component
Testing the application	test helps debugger	explanation component
Operation	interpreter	reasoning/ interference component with pre-defined solution strategy

Figure 1.2. Analogous Expressions of Application Programming and Expert Systems

2. Problems with the Programming of Multi-tool Cutting Lathes

Problems with the programming of cutting lathes mainly result from the extended treatment possibilities of these machines, which are multi-tool machinings (fig. 2.1), the use of rear drilling attachment as well as the use of driven tools for the C and Y axis machining (fig. 2.1). At multi-tool machinings the setting and modifying of synchronization points for the single tool carrier makes high requirements to programmer. On the one hand it is his task to guarantee an optimal proceeding of the parallel processings. On the other hand geometrical and technological collisions have to be avoided.

If rear drilling attachment treatments have to be made, a difficulty arises, i.e. that usually time-parallel treatments on the main and synchronous spindle have to be programmed. Another problem is that due to the process, instead of the tool the work piece is moved and correspondingly has to be programmed in another system of coordinates.

With the use of driven tools, for the C and Y axis treatment, the programmer has to fulfill the requirement to take another procedure process into consideration. In addition, the typical tasks of milling work have to be solved. Problems here are e.g. the consideration of further systems of coordinates as well as the programming of procedures in space. Also machine- and control unit-specific restrictions have to be taken into consideration.

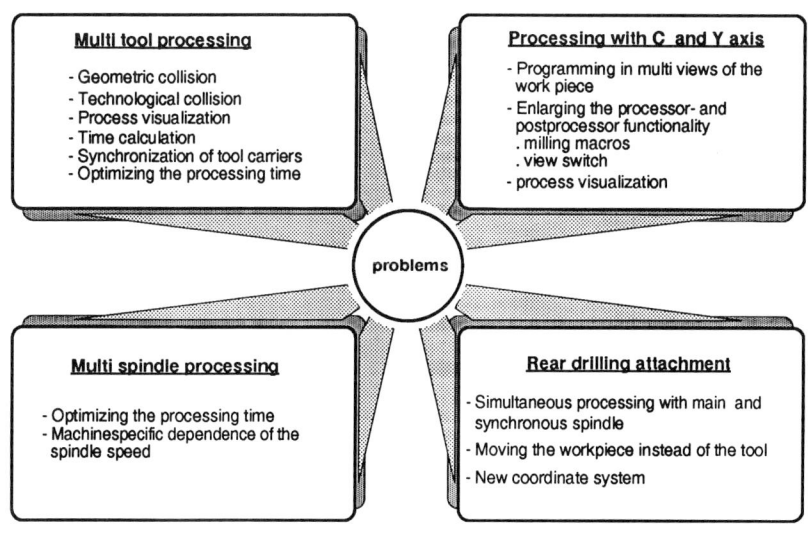

Figure 2.1. Problems with the Programming of Complex Lathes

3. Construction of a Programming System for Multi-tool Cutting Lathes

Refering to the problems described in the previous chapter, a general design for a programming system to support the user when dealing with complex turnings is introduced. As it is shown in figure 3.1 it provides different system components. On the one hand, they serve as an interface between user and programming system (interaction components), on the other hand, they contain method components and the data base/- administration as well as the system control.

The design of the programming system has different levels which are run during the drawing up of a NC programme. Below they are explained with the most essential system components in them. The sequence followed corresponds to the succession during the programming procedure.

At the beginning of the programming, the user can call up system components within the workpiece data preparation level for the CAD/NC coupling. They serve the purpose of taking over workpiece information from CAD systems. With the help of preparation functions the construction-orientated workpiece model is transfered into a manufacturing-orientated one /4,5/. Also algorithms for the production of a workpiece model are available as CAD functions. This CAD functionality should be so extensive that workshop or fixing drawings can be generated.

Within the part programme generation level the parallel editor with representation module as well as the possibility to use form features are essential functions in order to support the user when programming complex treatment tasks. The following chapter deals more precisely with the further automation of the part programming with the use of form features.

The parallel editor allows, contrary to the former sequential procedure, to build parallel running treatment sections parallelly and directly assigned to the respective tool carrier. Thus, the programmer has the possibility to influence the interaction of different tool carriers directly already when he is drawing up the programme.

The task of the representation module, which is tightly connected with the parallel editor, is to visualize adequately the sequences of different treatment sections which are partially time-parallel. Here, the user may call up different representation functions, like e.g. the synchronized representation of the part programme over all tool carriers. Via another function the treatment times can be represented in a shortened form. This is done in the form of a bar chart and states at the same time the charge of the single tool carriers (fig. 3.2).

At the part programme processing the NC sentences are generated which are necessary for the respective CNC machines. At the same time, there is a tight connection between part programme and NC programme sentences. This connection is necessary since the time calculation bases on NC sentences and the calculated process times are to be assigned to the original part programme.

The task of the function blocks assigned to the NC data processing is to reproduce particular functions of the control unit of the CNC machine. One has to notice that the number of the NC programmes, which are necessary for the processing of all the treatment operations of one fixing, can differ depending on the used control unit.

One of the function components of the NC data processing is an NC interpreter, which has to be in a position to take the interpretation possibilities of NC programmes, allowed according to DIN 66025, into consideration.

Another one serves the purpose of synchronizing the single tool carriers according to the NC programme. The module for the time calculation determines the proceedings of every tool carrier. It also takes machine-dependent times into consideration, like e.g. to switch single turrets. The modules for the time calculation, synchronization, geometry and technology processing strongly depend on each other and thus can not be regarded as individual modules.

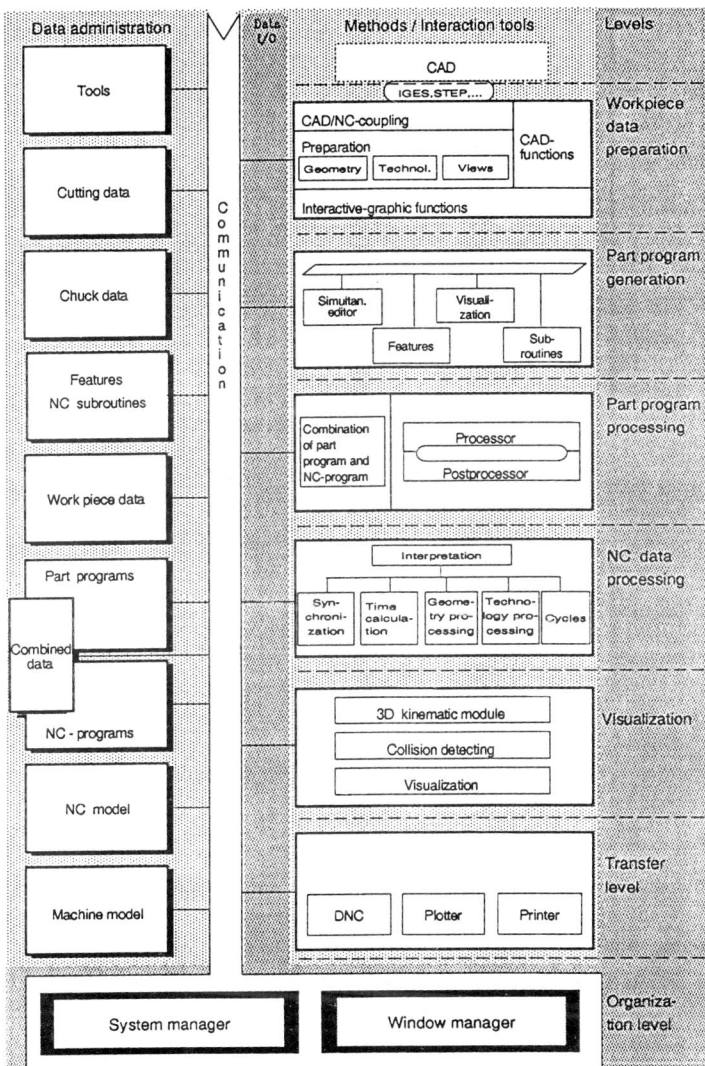

Figure 3.1. Construction of a Programming System for Complex Lathes

Within the drawing up and modification level, the results of the time calculation are reused by the representation module of the parallel editor as well as of the geometric treatment simulation by the modules for collision calculation and for graphic simulation. For the user, collision calculation and simulation are an important aid in order to check complex treatment programmes. They base on a 3D machine model. The movements of the modelled machine elements are generated by a 3D kinematic modul and can at the same time be reused to update the workpiece model.

Within the transfer level, the output of the generated and checked NC programmes is done, according to the application, in different formats and on different media.

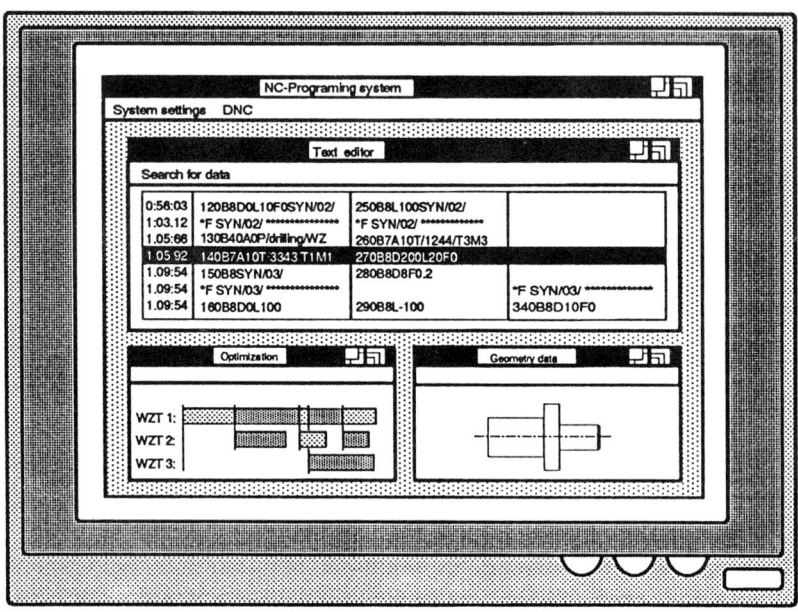

Figure 3.2. Parallel Editor

In order to support the programming, it is important that each sentence of the part programme is processed by the programming system right after the input. Here, particularly the synchronization as well as the carrying out of the time calculation is meant, in order to explain the user the effects of the sentence put in. A requirement for it is the possibility to be always in a position to change between the different levels of the programming system. The organization of these changes is the central task of the system control within the organization level.

There is a tight connection between these changes, which are often due to the proceeding, and the task of the window manager who carries out the control and administration of these windows, called programming windows. Another task of the system control results from the data transfer between the system components of the single levels and the different data administrations.

This communication is not the interaction of the user with the programming system via the user surface, but the data transfer between single modules of the programming system.

4. Use of Form Features at the Workpiece Data Preparation and the Part Programme Generation with the Help of Knowledge-based Methods

For complex turnings a further automation of the programming is required. This aim can be achieved among others by using knowledge-based methods. Thus, within the programming system level of work piece data preparation and even stronger within the level of part programming there are manifold operational areas as there exists, in comparison to the other levels, a high amount of heuristic knowledge which is usually replicable in the form of rules and methods. Another possibility is to work object-orientated via form features. In the following, a procedure for the preparation of the workpiece data and the generation of a part programme on the basis of form features is described (fig. 4.1).

The incoming information for the workpiece data preparation level is represented by the product model including all information on the workpiece to be produced. Within the form feature analysis it has to be found out what a workpiece is assembled with /6/. Another task is to transfer construction-orientated form features into manufacturing-orientated ones. This is due to the fact that often there is no 1:1 transformation of drawn elements into machining elements, like e.g. fins for the constructive reinforcement which have to be regarded as a pocket. The form features which have been determined within the form feature analysis can usually be subdivided and are thus called macro form features.

In the following, the analysis of the macro form features according to manufacturing aspects, which is the first task of the part programming level, will be further dealt with. Here, the macro form features like e.g. 'drilling' have to be subdivided into elementary forms (EFE) like 'centering', 'predrilling' and 'drilling'.

On the basis of geometric aspects, like e.g. lengths or diameter, manufacturing-orientated EFE have to be found out at which a succession suitable for the production has to be respected.

The knowledge required for this task can be represented in two knowledge bases and can be worked with the help of the inference component (fig. 4.2). The first knowledge base contains, as a workpiece model, all information about the piece to be produced. In the second one, which contains the process model, all information which is necessary for the analysis of the macro form features of a concrete workpiece is comprised. The knowledge included in the workpiece and process model is represented with the help of semantic networks and frames.

On the basis of geometric aspects, like e.g. lengths or diameter, manufacturing-orientated EFE have to be found out at which a succession suitable for the production has to be respected.

The knowledge required for this task can be represented in two knowledge bases and can be worked with the help of the inference component (fig. 4.2). The first knowledge base contains, as a workpiece model, all information about the piece to be produced. In the second one, which contains the process model, all information which is necessary for the analysis of the macro form features of a concrete workpiece is comprised. The knowledge included in the workpiece and process model is represented with the help of semantic networks and frames.

A semantic net is a relational connection of objects. The objects are treated as knots, their relations as edges. An object can be described with a frame. A frame contains boxes called slots for all the information connected with the object, like e.g. the drilling length or diameter.
These slots can store data or pointers to other frames, groups of rules as well as procedures with the help of which data can be calculated. The advantage is that objects having the same attributes can be comprised in classes. Then attributes simply have to be related to the class knots. A transmission mechanism cares for the passing-on of the attributes to all the class members. If the attributes of an object are related to specific data (e.g. diameter 1 = 10.5 mm) the fact will be called instantiation.

Along with the rules there are also methods within the knowledge base of the process model; thus one can not simply talk about a mere rule-based procedure. While rules describe condition-bound actions (if..., then...), a method is the procedure necessary to solve a certain task. An example would be the method to produce a tapping.

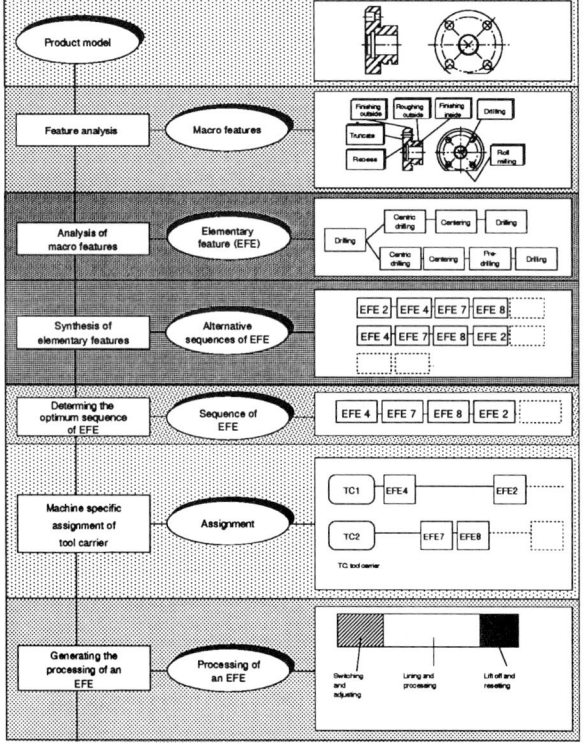

Figure 4.1. Use of Form Features at the Workpiece Data Preparation and Part Programming (EFE ... Elementary Form Features)

The proceeding of the knowledge contained in both knowledge bases is done with the help of the inference component by drawing conclusions. This is a process in which, controlled by an operation control, out of existing knowledge and existing assumptions respectively new knowledge and further assumptions are gained. In this connection, new means now available, previously not directly available. The mainly known inference strategies and operation controls are shown in fig. 4.3. Mainly, the connection mechanisms and search strategies are essential for the given task.

Alternative treatment sequences for the complete treatment task are the result of the ensuing synthesis of the elementary form features /7/. For this purpose, all the treatment alternatives which are possible for the production are set up and, according to the extended treatment possibilities of modern cutting lathes, different procedures like turning, drilling and milling have to be taken into consideration. Usually a great number of EFEs results form the complex turning tasks, thus there is a big area to search in and accordingly many solutions.

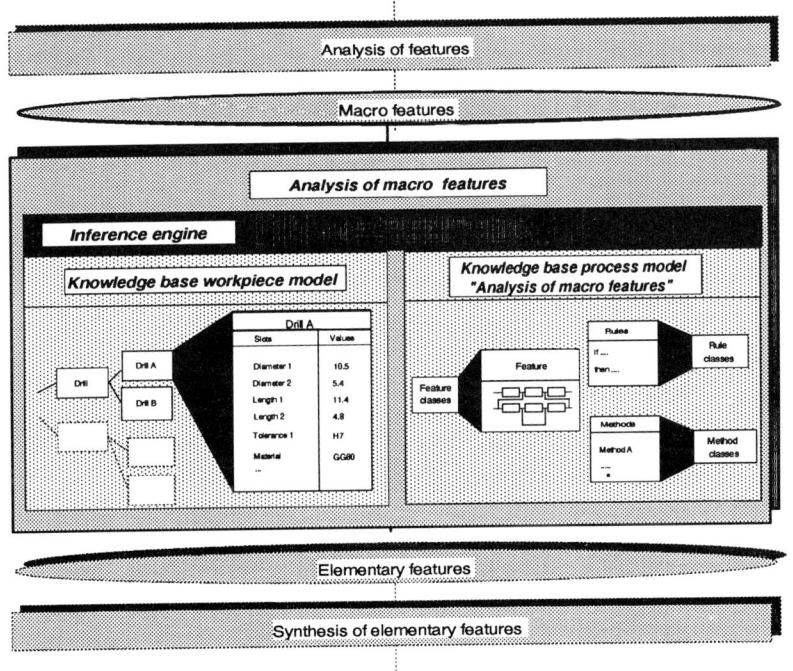

Figure 4.2. Knowledge Bases for the Analysis of the Macro Form Elements

The optimal treatment sequence has to be determined by the number of alternative possibilities. Criteria for the choice for turning lathes with one tool carrier are among others the number of tools in use, the number of tool changes as well as the travel. For the machinespecific assignment of the tool carrier these are also essential marginal conditions to be taken into consideration. With turning lathes, which can use several tools at the same time, the number and the construction of the single tool carrier have to be taken into consideration in addition.

Subsequently the processing of the single EFEs is generated. The input parametres required for the calculation are, among others, the rated and the actual raw geometry of the EFE, data about the workpiece geometry, information about adjacent EFE's as well as clamping elements and tool data. The processing of the calculated EFE can be divided into switching and adjusting, lift off and return movements during which the tool is not at use as well as the cutting feed and treatment movements. This division is of great importance for the synchronization of the EFE's since, with the help of this, the amount of time-parallel machine movements can be increased compared to the total treatment time.

Mechanism of evaluation	Mechanism of connection	Seeking strategies	Reasoning type	Processing of fuzzy knowledge
Modus ponens $A \rightarrow B$ $\frac{A}{B}$	$A \rightarrow B, B \rightarrow C,$ $C \rightarrow D$ - Forward chaining	- Depth-First:	- monotone interference	- Factors of confidence: probability of the accuracy of a derivation or the validity of a derivation rule
	- Backward chaining	- Breadth-First	- nonmonotone interference	- Fuzzi logic statement: "X is a big number" 0<X<100 with Wk 0,1 100<X<1000 with Wk 0,3 X>1000 with Wk 0,6
				Wk...Probability

Figure 4.3. Inference Strategies

The cooperation between the single tasks represented can be controlled by a so-called blackboard architecture. Each knowledge base tansfers its relevant information to a common work store, called blackboard. A superior control programme constantly checks the blackboard and determines the next step.

The use of form features within the workpiece data preparation and the part programming is divided into the functionally different tasks shown. It shows that each of these tasks requires a differentiated procedure, and therefore a task-orientated use of knowledge-based methods is appropriate. The existing requirement for a possibility of an integration into the total concept presented in order to avoid island solutions will nevertheless be fulfilled.

5. Summary

During the last few years multi-tool cutting lathes have been developed, mainly aiming at complete treatment in order to achieve short order times and thus short manufacturing and delivery times. Problems arising with the programming of these lathes because of the increased treatment possibilities, i.e. multi-tool treatment as well as the use of driven tools, were shown. From this point of view, a general construction of a programming system was set up.

A characteristic feature of the programming system presented is the division into different levels, each containing user-orientated system components. On the one hand, they serve as an interface between user and programming system (interaction components), on the other hand, they contain method components, the data base/- administration as well as the system control. This statement is generally valid and can be made on the basis of existing NC programming systems by using existing system functionalities of the same kind as they are in NC processors and postprocessors.

Finally, it was shown that within the programming system levels workpiece data preparation and part programming manifold fields of application for the use of knowledge-based methods for the further automation of the programming of complex turning tasks arise. Therefore, functionally different tasks for the part programming with form features were defined. It can be seen that each of these tasks requires a differentiated procedure, and an user-orientated use of knowledge-based methods is necessary.

6. References

1. R. Lederer, R. Holderle, Zeitgewinn durch Komplettbearbeitung. mav 12 (1990), p. 30-31.
2. G. Jascht, Komplettbearbeitung auf CNC-Drehmaschinen. Werkstatt und Betrieb 122 (1989) No. 8, p. 261-624.
3. W. Rüter, Simultandrehen und Werkstückkomplettbearbeitung. tz für Metallbearbeitung 80 (1986) No. 6, p. 27-33.
4. W. Walter, W. Hofmeister, Universeller CAD/NC-Kopplungsbaustein für NC-Programmiersysteme. wt.-Z.ind.Fertig. 77 (1987) No. 3, p. 129-133.
5. W. Hofmeister, Th. Reibetanz, Toleranzverrechnung im Rahmen der CAD/NC-Programmiersystemkopplung. Ind.-Anz. 111 (1989) No. 30, p. 54-57.
6. F.-L. Krause, Wissensverarbeitung für die rechnerunterstützte Produktgestaltung. ZwF 85 (1990), No. 3, p. 146-150.
7. H. Weber, H. Dürr, J. Steinmüller, NC-Daten wissensbasiert ermitteln. ZwF 85 (1990) No. 11, p. 572-575.

Process-planning for complex machining of rotational parts by means of knowledge-based methods.

Prof. D. Kochan, Dr.-Ing. M. Hess,
Dr.paed. J. Oelschlegel, Dipl.-Ing. Vogel

Technische Universität Dresden, Fakultät für Maschinenwesen, Institut für Fertigungsinformatik; Mommsenstraße 13; o 8027 Dresden, Germany

Abstract

The new machining centers for turning allow a much more productive manufacturing in comparision with traditio-nal NC-laths. But for a best possibel utilization it is necessary to have intelligent software support. The paper deals with the technological and methodical aspects for software-extension in connection with 2x2 axis turning operations.

1. Introduction

The increased performance of powerful CNC-controllers is the basis for the development of new generations of CNC-machining centers also for rotational parts. This new kind of machining centers allows complete machining of turning, milling, drilling and other processes in one clamping position. Such extended possibilities require new approaches for NC-programming and the entire user-support. Especially the 2x2 axis controlled CNC-lathes with 2 independently controlled turrets allow increased productivity, but require much more expenses for programming in connec-tion with the full utilization of the given new possibilities. This requirements can be fulfilled by new specific processor functions including knowledge based methods.

2. Machine tool concepts and principles

The machining centers for combined turning, milling and other operations are characterized by:

- 2 turrets with independend tool-holder-system
- one ore more cross-turret(s)
- driven tools
- interpolation between x-and c-axis
- second main-spindle and auxiliary spindle

Independently of this joint feature there are usually different constructive principles (see fig. 1)

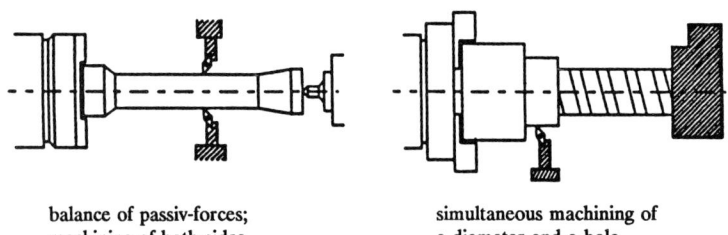

balance of passiv-forces;　　　simultaneous machining of
machining of both sides　　　　a diameter and a hole

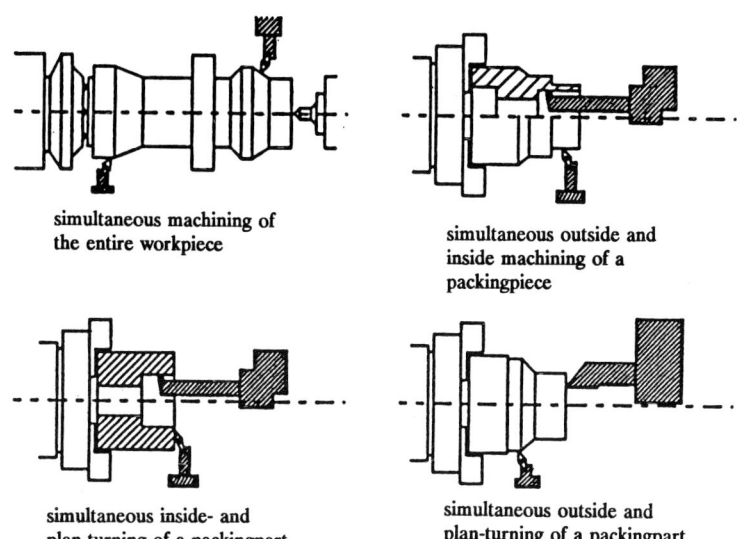

Fig. 1   two tools in operation (examples)

2.1. Constructive features and technological capacities of CNC-turning-cells

Essential developing-steps for turning-operations are given by the increased level of automation and equip-ment. Based on powerful CNC-controllers the available machine tool concepts are extended by modular struc-tures. For more then 50 % of all rotational workpieces there are additional operation necessary such as mil-ling, non-centrical boring and etc. /1/. A solution for complete machining are turning cells with driven tools. Typical for the current development is the broad application of CNC-turning cells with two independent programmable turrets.

2.2. Princips of machining-operations

By application of two simultaneously operating turrets the following machining-operations can be realized:

a) two tools in interference (fig. 1)
- simultaneous machining of one machining-segment (balance of passive forces)

b) one tool in interference
- turret two is used instead of the tailstock in case of long workpieces.
- turret two is used for workpiece-handling operations
- one toolplace of turret 2 is equipped with a driven chuck and can be used as a second spindle.

The realized constructions of machine tools allow the universal application for all kind of rotational work-pieces (shafts-, crank-shafts, lining-shafts...). Typi-cal variants of the arrangement of both turrets are de-monstrated in fig. 2.

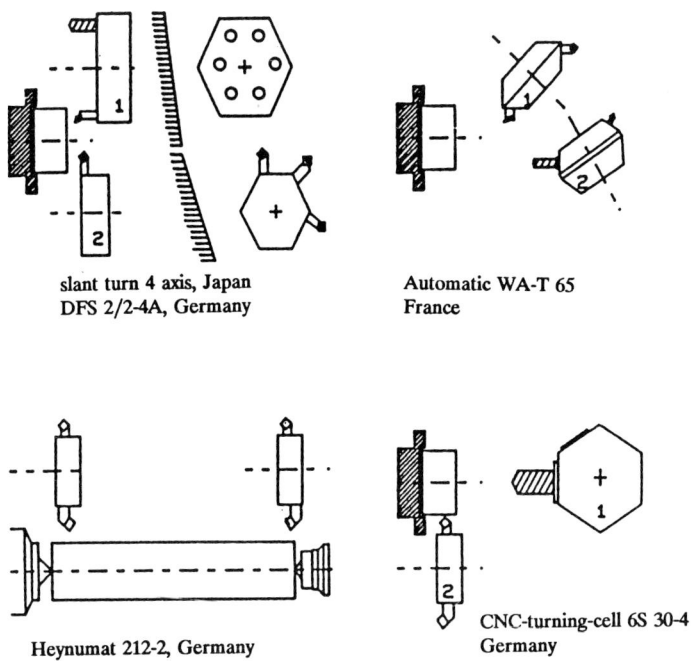

Fig. 2   different arrangement of turrets at turning-cells with two turrets

3. Specific requirements and problems

The extended machining capacities leads to higher expenses. This fact required high utilization of the installed equipment. In many cases only the advantage of complete machining in one clamping position will be used. This could be sufficient for relatively small parts. For the machining of shafts and other bigger parts it is necessary to provide the following conditions:

1. Advantage of the given technological facilities has to be taken by:
- a high degree of simultaneous machining into direction of minimization of machining time,
- complete machining for cost saving of the following operations.

2. A high technological reliability is required.

3. Reduced costs for the manufacturing preparation.

Definition:

As simultaneous machining we understand such machining-operations, where both turrets with tools are in interference including the necessary movement-elements for beginnings (run-up) and run-out.

To fulfill these requirements and conditions it is necessary to solve the following problems in connection with the NC-programming:

1. Evaluation of the technological compatibility at simultanious machining from the aspects of:

- quality assurance of parts (tolerances, surfacequality),
- safety of processes (stability of the system: machine tool-chuck-workpiece-tool(s)),

- complex relationships by cuttingforces, vibrations, chips between the operations.

2. Determination of optimized working values with consideration to:
- guarantee of the maximum possible rotary force;
- constant speed ($n1=n2=n$);

3. Realization of time and local synchronisation of all tool-movements with regard to:

- avoidance of collisions,
- observation of constrained machining procedures,

These complex problems can be solved only by support of specific programming systems or modules. Therefore, in some NC-programming systems these requirements will be fulfilled by the extension of available systems. In most cases only the assortment of movement-cycles of both turrets will be realized on the CLDATA-level in connection with time synchronisation and graphical simulation. The attainable degree of simultanious machining in these cases is normally very low. Both turrets are more or less used as toolstorage only. Some features of selected examples are demonstrated in fig. 1.

4. Characteristic of given development-levels

The fulfilment of user-requirements can be characteri-zed in three levels.

1. Concept "NC-editor"
2. Assortment in the processor, divided in the postprocessor
3. Processorextension as the highest level

1. Concept "NC-editor"

Main features of this level are:

- programming will be supported by graphical aids, in some cases by providing geometrical data from CAD-Files;
- usage of libraries or data-files for tool, working values...;
- application of graphical simulation.

So it can be pointed out, that the only advantage is the graphical support. All essential problems have to be solved by human knowledge and experience. Disadvan-tages are:
- On this level it is impossible to have objective technologi-cal evaluations.
- The determination of the working values is realized without consideration of the 2x2 axis specification.
- There doesn't exist any kind of computer aided support for decision making.

This solution is available on IBM-PC and is also partly realized in CNC-controllers.

2. Integration in the processor

This method is the typical one practically implemented at present. Main features are:

- relatively simple affiliation of cutting cycles to the turrets,
- simultaneous cutting of two different machining segments possible. In this case a reciprocal assortment of the cuts is possible.
- simulation and timecalculation after the postprocessor run.

The overlapping degree, which can be reached, is sufficient for small parts. For large and time-consuming parts the following disadvantages are to be stated:

- the actualization of geometric parameters with the aim of optimization of the cutting process is not possible;
- the determination of the workingvalues is realized without consideration of the 2x2 axis specific;
- the overlapping degree of operations will be relatively low.

This developing level is realized in some well known CAD/ CAM- and/or NC-programming systems, resp..

3. Processorextension

For large machine tools with time-consuming operation it is necessary to determine the optimum machining variant.

Main features are:
- subdivision of cutting cycles processor-internal,
- any specific assortment of cuts possible,
- actualization of geometric parameters into direction of minimization of waiting times possible,
- permanent checking of the time relations,
- finaladaption of turret movements by dialogue,
- graphical simulation.

These main features are realized in the system described in the following.

4. New concept and solution by means of processor extension

4.1. General principles

At the University of Technology a new approach, which fulfilled the above mentioned requirements was developed and implemented. Basis of the concept are the given variety of possible operation sequences. These operation sequences can be demonstrated by an operational graphchart. In this case the knots are the intermediate states and the edges are the operations or specific machining segments. The illustration the example of a "Driving-wheel".

The subdivision for machining into seven machining segments is possible. For the chosen manufacturing example there are theoretically 5 operation-sequences possible, but variant 5 is the usual one (overlapping of rough- and finish machining). Whereas at the two-axis machining the operation sequences are not of essential influence, the choice of the coupling of machining-segments and of operation sequences at the 2x2 axis turning is of high importance (fig. 3 and fig. 4).

Fig. 3 possible definition of machining-segments and demonstration of technical suitable machining-sequences for the example "Driving-wheel"

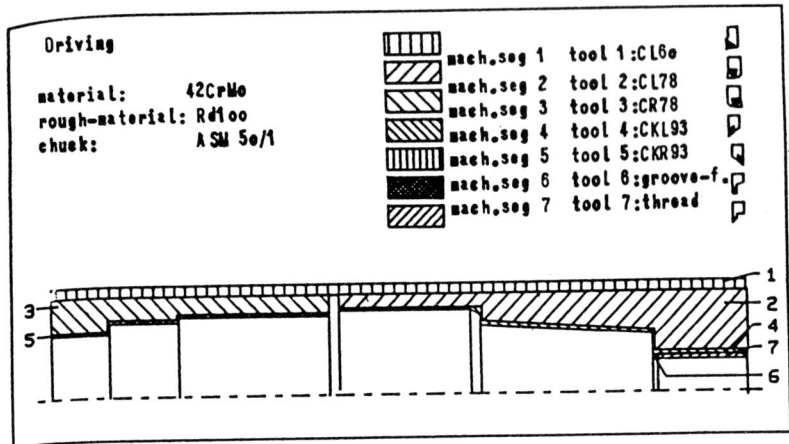

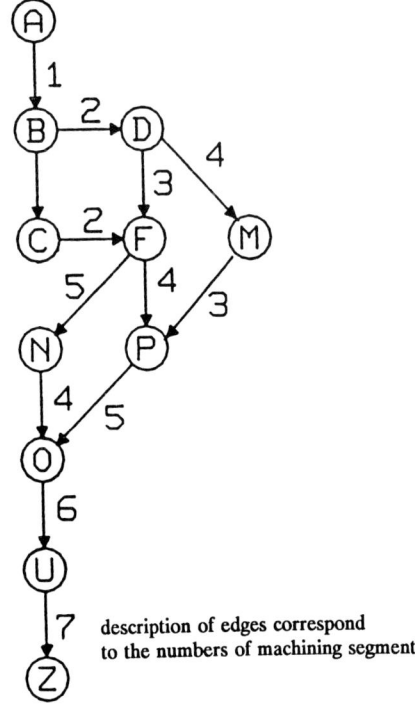

description of edges correspond to the numbers of machining segments

Fig. 4 demonstration of defined machining-segments and possible machinig-sequences for the manufacturing-example "Driving-wheel" for 2x2 axis controlled CNC-turning-cell

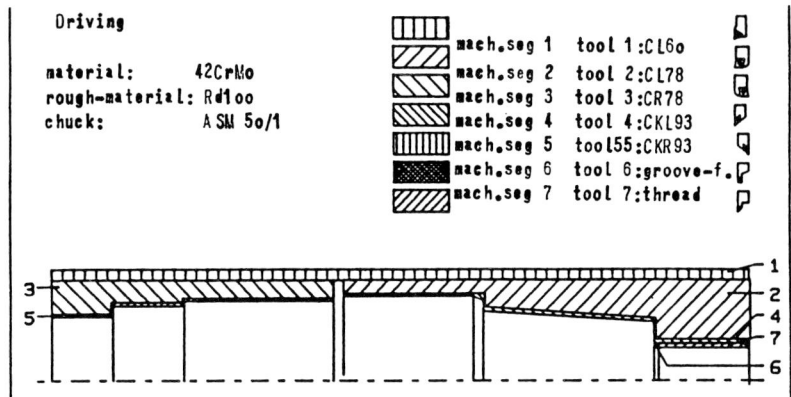

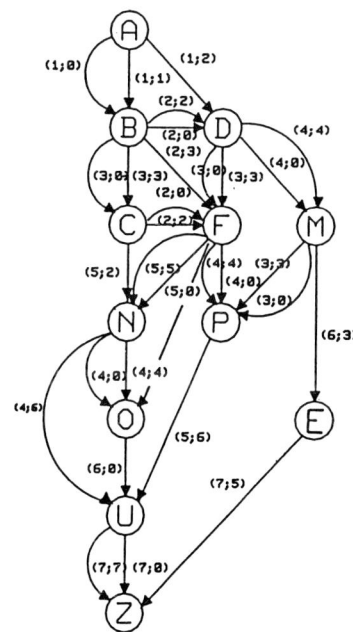

4.2. Performance

Usually the partition of the NC-control information to the NC-programming systems is realized on the basis of the CLDATA-interface. Contrary to this approach a processor-extension on the basis-programming-system called AUTOTECH-DR 43 was realized /2/. This leads to much more higher performance. Normally the manufacturing problem is given in form of a source code. As a result, from this startinformation two CNC-programmes, which fulfil the following requirements can be determined:

- timely synchronised,
- realization of collision-free machining,
- determination of workingvalues, which take into consideration the interference of the second tool.

The result of the userdialogue are manufacturing-documents, which guarantee by a high degree of simultaneous machining the effective and reliable application of these machine tools. Related to the example in fig. 2 in fig. 3. the time calculation for different variants of operation-sequences is demonstrated. It can be pointed out that the number of technologically possible variant increased rapidly. The selection and determination of economical operation-sequences is gaining in importance. Before the postprocessor run is calling in the possibility of graphical simulation of simultaneous machining will be provided. Both tools in interference can be demonstrated.

4.3. Methodical aspects

The efficient realization of the NC-language is possible by abstraction (transformation) into available soft-waretools. For supporting this approach powerful meta-systems like DEPOT 2a can be used /3/. The implementation and application of the NC-language can be realized in the following scheme:

	operation-sequence	Tgm	Tgm2A/Tgm4A
01	A-B-C-N-U-Z	10.18	84%
02	A-B-C-N-O-U-Z	9.53	78%
03	A-B-C-F-N-U-Z	8.73	72%
04	A-B-C-F-N-O-U-Z	8.08	66%
05	A-B-C-F-O-U-Z	7.89	65%
06	A-B-C-F-P-U-Z	8.77	71%
07	A-B-F-N-U-Z	8.86	73%
08	A-B-F-N-O-U-Z	8.21	67%
09	A-B-F-O-U-Z	8.02	66%
10	A-B-F-P-U-Z	8.90	73%
11	A-B-D-F-N-U-Z	8.73	72%
12	A-B-D-F-N-O-U-Z	8.08	66%
13	A-B-D-F-O-U-Z	7.89	65%
14	A-B-D-F-P-U-Z	8.77	71%
15	A-B-D-M-P-U-Z	9.05	74%
16	A-B-D-M-E-Z	7.89	65%
17	A-D-F-N-U-Z	8.14	67%
18	A-D-F-N-O-U-Z	7.49	61%
19	A-D-F-O-U-Z	7.3	60%
20	A-D-F-P-U-Z	8.18	67%
21	A-D-M-P-U-Z	8.18	67%
22	A-D-M-E-Z	8.72	82%

Fig. 5 time-calculation for different variants of operation-sequences

This method is used for the extension of the NC-programming-system AUTOTECH/DR 43 in connection with 2x2 axis controlled turning cells. The extended part of the NC-programming system is subdivided into the following module:

- automized variant determination of operation sequences;
- graphical dynamic simulation of workpiece machining;
- determination of working values ;
- loading of the tool holder.

For the generation of operating sequences a hierarchical rule-system was developed, which includes:

- experience based knowledge of the manufacturing engineer;
- simplification of sophisticated computing models for optimization;
- usage and enclosing of test results.

This rulesystem (approximately 250 rules) can be used for the evaluation of possible combinations of machining segments under technical and economical aspects. The result is offered to the NC-programmer for decision-making. Suitable combinations of machining segments which can be simultaneously machined are marked by blinking. Additionally the manufacturing engineer has the opportunity to explain of evaluations. The first level of the developed software has to consider mainly the following premisses:

- the rule knowledge is currently programmed in FORTRAN (in accordance with the entire NC-programming-system) This was possible in approximately two weeks. The disadvantage is given by the necessity to maintain the whole system by the developing engineer.

- the user-adapted notation of rule-knowledge and support by specific formalism (e.g. automatized hierarchy development) required approximately two men-years for developing the interpreter. In this case the NC-programmer is able to accumulate the experience based knowledge by himself into the rule base.

This it can be pointed out that the software-development is also a problem of optimization. It is not possible to find an ideal solution. The knowledge of the turning processes is prepared in a "preformalised" way. It resamble approximately a kind of decisiontables combined with graphical elements, which cannot directly be transformed into a computer aided mode. A logical analysis leads to the following conclusion. The rule system can be displayed in three levels:

- calculations of intermediate results, which can be further developed from fuzzy technological rules to not so fuzzy rules in dependence on cutting material and workpiece material,
- summarizing of rules from determined aspects of the particular scope (e.g. rough-machining),
- recognition of the hierarchical relations between rule groups.

An important problem in connection with the introduction of KI-tools in available systems is the question of data handling. For this requirement an interface was developed, which implemented the change of ASCII-data into facts of a knowledge base.

5. Summary

The extension of NC-programming systems for 2x2 axis machining required a lot of rules for technological and exonomical decision support. Such production and manufacturing rules can be prepared in kind of text files ore in graphical schemes. The main problem is the integration in available NC-programming systems. Our specific approach was implemented and succesful tested with some industrial examples.

Literature

/1/ BRÜMMER, U.; HESS, M.: Programmiersystem für 2x2 Achsen gesteuerte CNC-Drehmaschinen. Dissertation, TU Dresden, 1988;

/2/ Programmbeschreibung DR 43. FZW Karl-Marx-Stadt (Chemnitz) 1988;

/3/ GROSSMANN, R. u.a.: Depot 2a-Metasystem für die Analyse und Verarbeitung verbundener Fachsprachen. TU Dresden, WBZ MK/ IV, 1985/87;

Feedback for Process Planning as an Approach to Intelligent Quality Assurance.

F.-L. Krause, A. Ulbrich and R. Woll

Institut für Produktionsanlagen und Konstruktionstechnik, IPK - Berlin
Pascalstr. 8-9, D-1000 Berlin 10

Abstract
This paper represents an overview of a first realization of a feedback system that provides for the flow of comments from the shop floor back up to process planning. It addresses the notions of the functionality of feedback, of a feedback data model and a system concept as well as the application of feedback in updating knowledge bases for enduring improvement of product and process quality. If such an update process is integrated into already operational feedback loops, the described functionality becomes part of a self-learning system concept.

1. INTRODUCTION

Nowadays it is becoming ever more critical to shorten the lead as well as the throughput time for the planning and production of products. It is necessary to reduce the time required for making a product available on the market without increasing the cost of manufacturing. Alongside the vertical and horizontal integration of information flows, the feedback of experience gained in the production process on the shop floor back up to the design and planning process is also required to close the CIM-loops. To implement the design and planning imperative "do it right the first time," it is necessary to avoid loops to improve, modify or adapt drawings and process plans to requirements on the shop floor. The improvement, modification and adaptation of design and planning results is one essential component of quality assurance, especially if the use of feedback information is targeted at obtaining enduring and systematic improvements. Within the framework of the Esprit Project 2615 IMPPACT (Integrated Modelling of Products and Processes using Advanced Computer Technologies) the initial stage is being realized in the development of a mechanism for feedback of comments from the shop floor back up to process planning. Hereinafter, the following definition will be used:

Feedback constitutes the gathering of information about products during their entire life cycle and its use for correction and verification on every level of the production process. It represents knowledge about the product itself and corresponding production processes.

There are two points of emphasis to be considered: the short term aspect of improvement and modification of drawings and process plans based on shop floor information and the documentation as product history within the product model as one part of the product life cycle; and the long term aspect of the analysis of feedback information, e.g. the product history to derive new facts or rules to be integrated into the production knowledge data base.

The feedback mechanisms recently available provide support particularly in the areas of quality assurance and scheduling. A lot of data have already been gathered but they are not made available to the design and planning department in the appropriate way.

Frequently, the worker on the shop floor departs from a process or operation plan as a result of incorrect planning detected on the basis of his experience. Mechanisms are not provided to capture and report this information so that the designer or planner can take such factors into account later on.

2. ASPECTS OF FEEDBACK

In a complete feedback system exchange of information is made possible between all CIM-modules and on all production levels. A CIM-module can be regarded as the building blocks of a CIM-architecture. The modules can be integrated without an understanding of its internal structure. Only the functionality of the module and its interfaces are of importance. The module together with other CIM-modules can be integrated using data or procedural interfaces.

In this contribution emphasis was placed on feedback of information from shop floor to process and operation planning.

Four kinds of feedback can be distinguished:
- permanent feedback,
- special-requirement feedback,
- failure feedback and
- feedback without requirements.

Permanent feedback is used in running quality inspections or for shop floor control. The definition of collection requirements is produced once, and the whole collection, analysis and feedback process runs continuously within the production process. Much data is available today within the various special purpose systems, mostly in the form of private data bases on an aggregate level which offer nothing of utility in process planning. Measuring or Statistical Process Control data cannot be used easily by a planner directly.

Special-requirement feedback is found in prototype production. Planning results are available, but they have to be verified and adapted to the factory environment. Testing and verification of planning results is established as a common activity within mass or large series production. From a general standpoint it is useful to have the same mechanisms available in a small series or single part production environment. In this case a verification of planning results is done directly on the shop floor by the worker involved, because he has to do his job right the first time. Experience and information gathered here are of great interest in process planning. But at present the worker often makes modifications and adaptations without producing any feedback at all. A typical example is changing the NC statements at the machine tool without providing a mechanism for making information about these modifications available to the NC-programming department.

Failure feedback means a failure on the shop floor such as a tool break down requiring immediate action. Feedback concerning such events is crucial in the identification of especially obvious factors such as wrongly calculated process parameters.

Feedback without any requirement provides the worker on the shop floor with the opportunity to make comments, explain deviations and their noticeable results or direct improvement proposals upwards to the planning echelons. He should provide commentary concerning all changes he makes. This presupposes a personally cordial and technically "user-friendly" interface offering a welcome reception of such commentary and a high level of motivation on the part of the worker.

Worker motivation will be the real bottleneck for future implementation of feedback systems. The only technical solution is to make the feedback workstation and the whole environment on the shop floor as user-friendly as possible. A further technical possibility is to enhance communication facilities by means of integrated workstations on the shop floor. The worker can be motivated by being shown the effects of his optimization of processes on process planning. In this way he can directly experience his involvement in the flexible planning procedure through his use of feedback facilities.

The next aspect of interest to this investigation is the informational content of feedback. A process plan model contains "proposals" for the here relevant objects:
- **processes** for the generation of appropriate
- **product** characteristics such as features and the
- **factory** equipment required for these processes.

A deviation in the actual production process may occur for any object or its attribute. Deviations beyond the acceptable limits may be seen as failures. Such failures can be structured as in currently available quality inspection systems. Any given failure may have a number of causes and effects. These causes and effects may form the basis for other deviations. The network of relationships between all possible failures, including those resulting from the causes and effects of other failures, must be described in an analysis.

A combination of proposed process, equipment or product characteristic and possible failure is hereinafter called "feedback topic". Feedback topics are predefined for the procedure of feedback collection on the shop floor to simplify the input of comments coming from the workers and following evaluation procedures.

The advantages of predefined feedback topics are: Input of comments on the shop floor is simplified if an event is describable by predefined feedback topics. Hereinafter this information will be called "structured comments".

Structured comments provide for the possibility of computer aided evaluation of events, especially for the update of process knowledge bases.

The capability of submitting arbitrary feedback information will be provided in the form of the "free comment", which can be described as an unstructured description of a deviation within the framework of a feedback topic.

3. FUNCTIONALITY

The purpose of feedback is to bring the relevant aggregated information from the shop floor back to upstream systems. To implement the feedback procedure a system is required which contains the following modules:
- prepare feedback,
- collect information,
- analyze information and
- manage feedback.

Prepare feedback

The function 'prepare feedback' is useful for the determination of information collection requirements. If any CIM module needs feedback information it sends a message about its requirements to the feedback system. The first step is to display these requirements for review by the user. If a requirement is defined, a corresponding feedback topic should be available covering that requirement. Available topics will be shown on the screen. If necessary topics are not yet defined, they have to be added to the system. Perhaps required information has to be broken down into collectable data. This is a task for the user who prepares the feedback. The user adds and reviews requirements.

External to the system there are special application modules available such as
- inspection planning,
- programming of measurement machines or
- statistical process control preparations.

The ability of calling such modules will be available at a later development stage of the feedback system.

Collect information

Shop floor activities must be performed by the worker. In performing these activities, the worker has the opportunity to
- collect the required data,
- describe special events,
- submit comments regarding actual process changes or
- submit comments without a request.

In all of these four cases user is given the opportunity of adding and reviewing comments, but under different aspects. In the first case the required feedback topics are automatically presented at the workstation or terminal. In the second case a special list describing possible events such as a breakdown or a collision will be displayed and a rough description of obvious causes for these events should be given. In the third case the change of process parameters or of required tools or fixtures is described. The fourth case describes the situation where the worker wishes to submit proposals for optimizations or notifications concerning a frequently occurring failure such as incorrect cutting parameters. To carry out these collection processes in an efficient way a user-friendly interface which is
- easy to operate,
- clearly structured and
- flexible with respect to requirements in varying shop floor environments is needed.

Analyze information

According to the information collected, priority flags can be added to make sure that information describing important events such as a tool breakdown is processed as soon as possible. Of great importance to the effectiveness of the analysis is the ability to add effects and causes to the feedback topic. Submitted improvement proposals are the point of departure for an optimization in process planning. Based on functions for information aggregation, reports are generated which will be formalized in a feedback formsheet.

Based on particular calculation methods the data captured during the collection procedure will be analyzed and aggregated. That means to summarize information from different points of view and to represent them in the feedback archive in a compressed format, for example based on a set of measuring points indicating deviations between a planned and a realized part. The causes of the deviation are derived from an evaluation of the measurement information and stored in the archive as free-format textual information.

Manage feedback

This module is crucial mainly in closing feedback loops. In addition to feedback target determination mechanisms and the messaging procedure, facilities for management of feedback information are necessary.

Based on the analysis report, new feedback topics will be introduced or old ones will be dropped. "Free comments" must be transformed according to their similarity and frequency of occurrence into "structured comments" based on feedback topics.

The determination of the target CIM module for feedback requires a description of CIM module attributes as
- CIM module's name including the version,
- CIM module's location describing the department where it is used,
- CIM module's functions which are described by a verb and an object, such as 'select tool' or 'determine cutting data'

A comparison of the improvement proposals, comments, causes and effects associated with each feedback topic with the CIM module attributes enables the user to identify a specific CIM module or specific CIM modules as target(s) for the addition of the available feedback information. If, for example, a tool is broken during production caused by an excessive feed rate as a parameter of the cutting data, the first target for the feedback topic 'tool is broken' is a CIM module assigned the function 'select tools' for informational purposes, and the second target for the feedback topic 'feedrate too high' is a CIM module assigned the function 'determine cutting data' for improving the determination process.

The last function connected with comment processing is the transmission of a message to the target CIM module(s), display of improvements and addition of the improvement status to the feedback archive.

4. INFORMATION MODEL

The feedback data model is based on the requirements of the above mentioned functionality, additionally taking the characteristics of the CIM environment into account. The feedback module mentioned here is part of the IMPPACT system concept.

Within this concept several CIM modules communicate via an integrated product and production process model. The process plan representation is one part of this model. It contains product, process and equipment describing entities such as tools, cutting data or features and their attributes such as diameter, feed rate or spindle speed. The feedback model is supposed to describe deviations from the expected parameters. Deviations occurring on the shop floor can be related to an attribute of an entity (diameter to small, feed rate to high) or to the entity itself without a further specialization (tool broken, feature is missing).

For the general description of deviations the concept "characteristic" is here introduced, encompassing either an entity itself or an attribute of an entity. The combination of a characteristic with a deviation is hereinafter referred to as a feedback topic.

Possible deviations can be derived from a deviation type. The deviation type is directly dependent on the characteristics and can be structured in correspondence with them.

A deviation is described by two attributes:
- the deviation value, the difference between expected and actual values and
- the deviation description, which characterizes the deviation of an entity as a special failure.

All possible feedback topics can be described by means of these if a deviation type has been assigned to the characteristic, figure 1. During the subsequent feedback process only selected, predefined feedback topics are taken into consideration.

Another view of feedback information is related to the distinction between generic, specific and occurrence information. This construct is a general one out of the IMPPACT data model. It describes the specification procedure of a special kind of entity.

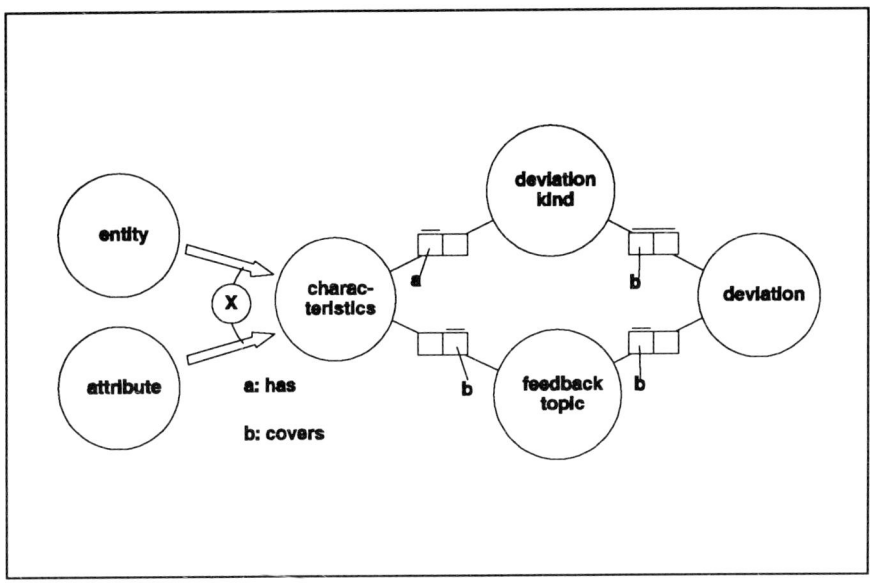

Figure 1. NIAM chart of the embedded feedback topic

Example:
A "generic cube" consists of defined attributes such as length, width and height of a cube whose values, however, need not be defined. If all of these attributes have fixed values it is a "specific cube". The cube can occur several times at different places and at any time in a predefined world, so it is called an "occurrence cube". It is an instantiation of something "specific" in the real world.

The specification procedure used on feedback topics results in the following constructs:

Generic feedback topic

This possible feedback topic is not fully specified. It should be predefined. At the minimum it possesses only relations to an identified entity in the process plan representation and a relation to its kind of deviation and its possible deviation description space. For example:

spindle speed   maybe   to high or o.k. or to low
tool   maybe   o.k. or broken or worn out.

All predefined generic feedback topics are stored in the generic part of the feedback data model. Additional deviation possibilities will be stored there as well in a structured way.

Specific feedback topic

Generic feedback topics are specified according to specific entities in the process plan representation. A specific feedback topic is linked to a specific entity, where all values for attributes are available and the deviation description is settled, for example:
spindle speed (of) 1000 rpm (is) too high.

Occurrence feedback topic

The specific feedback topic described above may occur several times in production in one or many different projects. It is then project dependent. Any occurring deviation of an entity within the process plan model is an occurrence feedback topic and has additional descriptive attributes such as
- occurrence date or
- observer name etc.

For the description of special situations which cannot be described by the predefined feedback topics, the free comment option is provided. Free comments are not structured. If similar free comments are entered many times they have to be analyzed by a feedback manager or analyzer. If required, additional feedback topics have to be inserted into the generic part of the feedback data model.

The basic model structure is shown in figure 2.

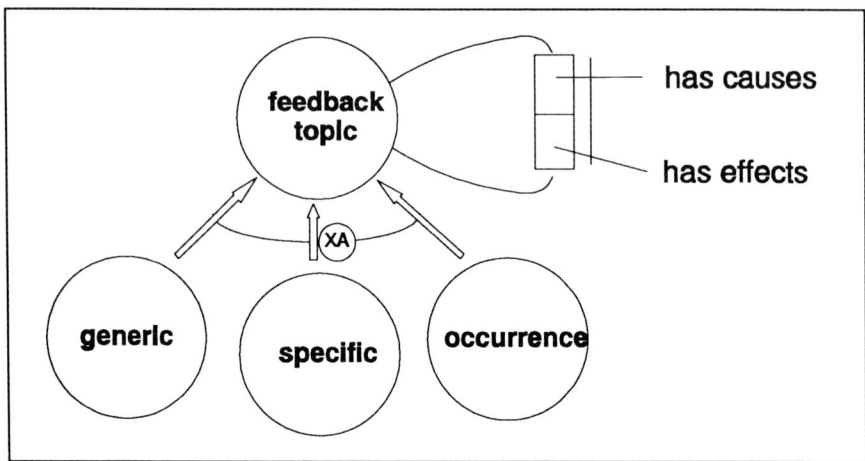

Figure 2. NIAM chart describing the basic construct of the feedback information model

To support the whole feedback process it is sensible to introduce additional entities to describe failure-effect networks. One feedback topic may be a cause of many of other problems as much as an effect of other problems. This description provides a basis for the definition of networks to any extent desired. In the following, this construct forms the point of departure for further analytical work. Priorities are linked to the effects of any given problem describing how serious an impact they have on production.

5. SYSTEM CONCEPT

The essential idea behind this concept is the transformation of process information from the shop floor into a feedback data base and subsequently into a process planning CIM module including a mechanism for updating CIM modules' knowledge bases.

The transformation process is described in general in Chapter 3. The four functions mentioned all interact using the central feedback archive, figure 3.

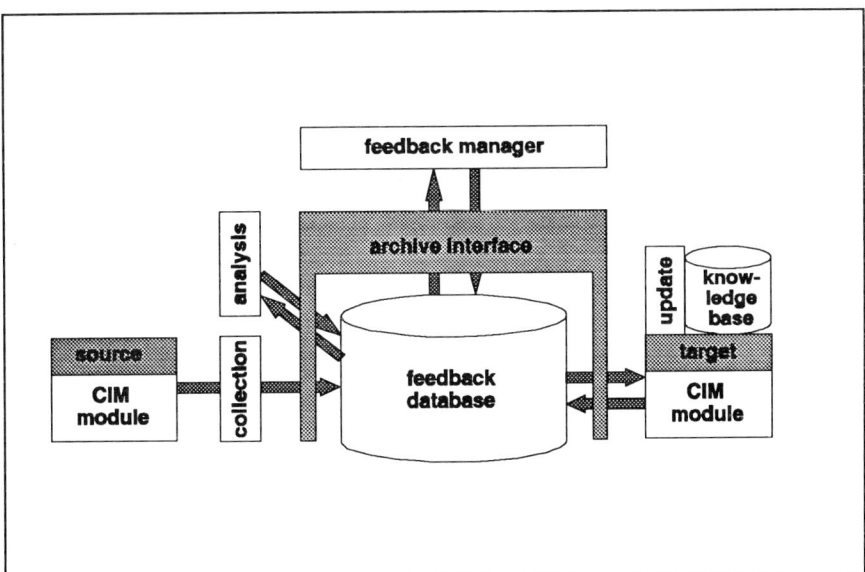

Figure 3. A feedback system concept

The ability to extract feedback information from this archive by means of other functions should be provided via an archive interface. All the archive information must be referenced to the product and process model in the central data base. The feedback archive is from a conceptual standpoint a logical part of this central data base.

The collection functionality needs a dedicated preparation to define information to collect. Preparation functionality is a part of the feedback manager. The collection solves the extraction of data and the input to the archive. There may be several analyzing packages available to
- aggregate data
- analyze possible targets
- derive special conclusions.

They all work in the same way. They get information out of the archive, process them by using own algorithms and store aggregated higher level information back in the archive.

The feedback manager controls the whole process and sends messages of available relevant feedback information to the distinguished CIM modules. He offers also a special library of prestructured feedback topics mainly for easy and defined data collecting. This library may be addressed as the generic predefined part of the feedback archive. Feedback information on different levels of aggregation may be sent to the target module if it is able to handle it.

6. USE OF FEEDBACK INFORMATION

Feedback information from the shop floor constitutes a starting point for significantly influencing the quality of products and processes.

Influences produce effects both short term and long term. Process plans can be adapted in the short term by making changes to entries and specifications in the relevant documents. This results in a quality improvement as a consequence of reuse of the changed documents.

In the case of a newly produced process plan, however, the same errors are produced as long as the underlying elements of the planning process are not effectively changed on the basis of the feedback information. This requires a change in the knowledge base. The necessary procedure is described below.

By an analysis of feedback information, stored in the feedback data base, knowledge acquisition can be carried out. This will be done by interpretation of the feedback history concerning specific feedback topics using feedback handling functionality as well as basic analysis methods. The result for example is a sheet of paper describing deviations of planned and realized values for specific operations corresponding to one production feature, which happened frequently.

By interpretation of these results incorrect data, rules or behavior of equipment can be determined and corresponding modifications can be carried out, for example corrections of cutting data, basic times, machine tool accuracies or wear and tear of clamping elements as well as rules for tool selection or operation sequencing

By updating the process knowledge based on this acquisition the feedback loop will be closed. That leads to an enhancement of the intelligence of the system and the quality of the planning results. Therefore configuration components are required, allowing the integration and update of process knowledge consisting of planning data and rules represented in the process model.

Process knowledge can be structure corresponding to its representation in the following way:
- Resource model information which is represented in data bases such as cutting data, times, material data or equipment data,
- General process knowledge which is represented in an object oriented way providing basic methods, e.g. for the determination of operation sequences, machining strategies or for the selection of tools or clamping devices,
- Company specific knowledge, highly dependent on resources as well as product type, to be represented by rules.

Knowledge modification and integration will be carried out on the one hand by updating the resource information in the data base system and on the other hand by editing rules. Both procedures make it possible to influence planning results without any modification of the software itself.

Essential for a process planning system is the integration and modification of rules. Rules establish the link between production features and the corresponding machining methods for the generation of machining information such as cutting strategies, tools, fixtures and operation sequences.

To carry out company or product specific process knowledge adaptions easily, functions have to be available enabling the definition and manipulation of rules. Based on an implicit feature description such as dimensional parameters or tolerances, the corresponding process descriptions are attached to the production features as rules. The process rules belonging to one specific feature are structured in rule classes related to the task they have to fulfill, e.g. determination of processes or operations, selection of tools or machine tools, determination of clamping or cutting strategies.

The objective is to provide a system component, supporting knowledge and data processing in a flexible way without recompiling and relinking the whole planning system and without development of new software components for knowledge integration. This requires concepts allowing the user to define company specific rules without any experience in software development. The rule definition component must be independent of the rule processing mechanism. Therefore a compiler will be used, translating the rules from an external representation into a system internal syntax. For rule processing itself AI-methods providing inference mechanisms are used. For integration of a rule processing component, interfaces to other programming languages or system components such as data base systems or user-interfaces have to be available. It has to be possible to invoke the rule processing component using methods written in object oriented programming languages such as C++ and to transmit objects. A further requirement is the ability to call functions from the rule processing component. Another requirement is an interface to a data base system where the facts the rules are working on are stored, to extract only facts relevant to the task to be carried out. The results of a planning procedure can be influenced on the one hand by modifying the facts in the data base and on the other hand by modifying the rules directly.

Once the knowledge has been updated, the new decision rules will influence the future planning process, so that careful verification of updates is necessary. In general there are two distinct options for verification:

- performance of the verification process within the module affected by
  . testing the updated knowledge base on former projects and analysing the new results for effectiveness, or
  . using truth maintenance facilities;
- sending a message to the feedback system to dispatch a special requirement feedback transaction for a new project using the process plan generated based on the updated knowledge.

A complete feedback loop can thus be established for optimizing planning results in parallel with the current planning task.

The first step in the implementation of the knowledge update facility is the realization of a component for updating equipment models within the feedback process. The equipment archive is extended by additional functionality for updating equipment data within the scope of feedback processing. Communication with the feedback module is accomplished via a mailbox system which dispatches messages triggering an update of equipment data. Connection to the feedback data model will be established using the data base interface. Based on comments from the shop floor concerning usage of equipment, information about wear-and-tear conditions or changes in precision is attached to the components involved for use in process planning. This feedback mechanism leads to improvements in future planning results.

7. FUTURE PROSPECTS

Feedback functionality for comment processing as described above in Chapter 3 provides the conceptual basis for the future development of systems which deal with or require comments describing production performance. The realization of a first prototype will demonstrate the possibilities of feedback; the prototype itself could, in the future, find direct application in any area of production or design as well.

User friendly feedback systems must in the future be constructed around powerful graphic user interfaces providing a realistic, graphic representation of problems occuring in the production process, as well as for the capability of creating a visually representable link between feedback information and the "product gestalt". This opens new possibilities for quality control by offering graphical representation of failure networks.

Further expansion of the application area of feedback processes is necessary, for example, to integrate feedback about usage, recycling and disposal. The feedback functionality has to be one component part of an integrated quality control mechanism within product development. Data links defining the applicability of feedback information must be extended beyond factory walls in integrating suppliers and sub-suppliers into the data model representing the entire product life cycle. A future-oriented application of feedback mechanisms is made possible when central knowledge bases are available which offer knowledge continuously brought up to date by adaptation on the basis of information obtained via feedback about actual production conditions.

The integration of such a functionality into overall CIM structures is a further task in any future-oriented development. It will be a crucial aspect in the realization of simultaneous engineering. Particularly the feedback of information between any given level and all other possible levels in the production process must be supported, for example between design and process planning or NC-simulation and operation planning. To reach this goal, the feedback concept must be further developed to provide corresponding functionality in the areas of handling and monitoring. An additional important aspect is the increasing integration of methods for automated data capture on the shop floor.

Verification of knowledge base in intelligent technological design system

Kolchin A.F., Zykova S.A.

CAD Department, Mosstankin, Moscow, USSR

Abstract

The presented paper deals with the problem of knowledge verification in intelligent technological design systems. A vital need for building up a knowledge analysis block is substantiated, a logical model of intelligent system based on many-sorted theory of resolutional type is outlined and notions related to knowledge base verification are discussed in detail. In conclusion a table, containing the main errors, occuring in knowledge bases, methods of their detection and correction, is given.

INTRODUCTION

The utilization of methods and means of an artificial intelligence while solving modern technological design tasks becomes vitally important. In particular, investigation of general principles of the creating of the system supporting the technological process development for manufacturing parts of rotation type leads to the concept of intelligent system with elements of self-training, a possibility of the modification and self-modification based on the information received and accumulated experience processing [1]. A number of factors which are encountered by designers of modern intelligent systems brings about an acube problem of support and rational utilization of knowledge bases, used by the systems of that kind.

One of these problems solutions is a creation of an independent superstructure above a standard data base - data analysis block; its purpose is analysis of knowledge contained in the base.

In this paper a vital need for building up a knowledge analysis block is substantiated, a logical model of intelligent system based on many-sorted theory of resolutional type is outlined and notions related to knowledge base verification are discussed in detail. In conclusion a table, containing the main errors, occuring in knowledge bases, methods of their detection and correction, is given. The results are based on the experience of the creation of the intelligent system, supporting the technological process development for manufacturing parts of rotation type.

1. GENERAL NOTIONS

By the term data analysis we understand a whole set of operations related to validation (to conduct validation means to show that the system is performing those functions which it is assigned to execute), rationalization (rationalization - the forming of formal knowledge structures in the most rational way from the point of view of their further utilization) and classification (for example, according to abstraction levels), knowledge verification. We shall be interested in knowledge verification problems.

Knowledge verification is an investigation of knowledge with the aim of checking its accordance to certain properties and demands, namely, the properties and demands of knowledge
- non-redundancy,
- consistency,
- completeness.

Detection of redundancy, inconsistency and incompleteness, localization of their sources and a corresponding knowledge base correction (automatic, if mechanisms of automatic correction are provided for the given detected situation) are the results of the verification process.

We shall differentiate between verification in case of a static knowledge base, that is investigation for redundancy, inconsistency and incompleteness of a knowledge base, the creation of which has been already completed (or some step of its creation was terminated) and verification during addition, removal, modification of some portion of knowledge, that is investigation for redundancy, inconsistency and incompleteness of dynamically changing knowledge base and new (updated) portions of knowledge.

Before passing on to the discussion of knowledge model and concrete verification stages, let us dwell on possible sources of errors, the detection and the correction of which are the aims of the verification process.

Errors can occur:
- in the process of direct input of information into the knowledge base by the user;
- in the process of receiving requests to the knowledge base directly from the user;
- in the process of entering the information, received with the self-training mechanisms utilization;
- in the process of reasoning - the solution of the task and complementing some knowledge portions to the data base, as well as available knowledge modification and removal.

The errors can also occur because of the shortcomings of the domain area investigation.

This is far from comprehensive list of possible sources of errors. Errors born can be of different character (for example, misprints, logical errors, gaps in a description of knowledge about the domain area) and can have a different degree of seriousness (for example, they can bring about the decrease of the system operation efficiency or the complete paralyzation or erroneous functioning). Specific classes of errors, detected in the process of verification, will be described below.

2. INTELLIGENT SYSTEM MODEL

2.1. Knowledge model

We shall stick to a concept in accordance to which an object area can be represented as a hierarchic set of data, information and knowledge [4].

The mathematical logic means are used for simulating an object area in intelligent system knowledge base. Logical means provide for an adequate representation of knowledge about the domain area, are commonly accepted formality and are closely related to standard data bases; logical knowledge representation can be easily modified in series of representations, based on other metaphors.

Knowledge about the domain area is simulated by means of MLH language, which is a language of many-sorted first order predicate logic [2,3], in which a class of perfect formulae is limited by a class of Horn clauses [2]. We shall identify a knowledge base with a set of Horn clauses. Let us establish a correlation between the domain area objects and logical model objects:

data set	---->	interpretatioin area
data unit	---->	interpretation area element (constant, variable)
data population	---->	sort
information unit	---->	tuple of elements of interpretation area (a copy of relation)
relationship	---->	predicate (atom), function
knowledge unit	---->	Horn clause
a group of knowledge	---->	a group of Horn clauses with a common main predicate
chain of inference	---->	a group of Horn clauses, composed in accordance with the principle of logical consequence.

We shall further on consider that all variables in atoms, comprising Horn clauses, are bound by universal quantifiers and we shall put down Horn clauses using replication, substituting conjunction by a comma:

B<---A1,...,An (1)
B<--- (2)
<---A1,...,An (3)

Clauses of the type (1) are called rules, clauses of the type (2) - facts, clauses of the type (3) - questions (requests).

Atom (predicate) B is called the main predicate of the clause, head or conclusion, atoms (predicates) A1,...,An - premises, and their set - a clause body.

The selected domain area interpretation presupposes the utilization of many-sorted peredicate logic for model build up. A many-sorted predicate logic can be embedded in one-sorted predicate logic by adding unary predicate symbols for different domain area sorts of the domain elements, $S^{\delta}(x)$, expressing that x is of sort δ. We shall make use of the possibility of this reduction and we shall work within the framework of first order predicate logic.

Limitations, imposed upon the class of permissible formular, decreasing the expressive power of a model, however, allow formalization of a wide class of knowledge, making it possible to use more effective and less labour-consuming inference strategies, compared to an unlimited case of the first order predicate logic.

2.2. Intelligent systems and many-sorted theories of resolutional type

Intelligent system is the system of knowledge representation and processing.

Under the above interpretation, intelligent system with knowledge base represented by set of Horn clauses can be considered as a theory of resolutional type in first order many-sorted predicate logic, namely, theory based on:

1) language MLH, i.e., language of many-sorted first order predicate logic, in which a class of perfect formulae is limited by a class of Horn clauses;
2) set of inference rules:
- linear input resolution,
- unification,
- rearrangement of premises,
- removal of the duplicating premises;
3) set of individual axioms, which is the set of Horn clauses, representing the knowledge base of given intelligent system.

The notions of logical consistency and completeness can be transferred correctly for the intelligent systems; because of the fact, that the problems of consistency and completeness of a theory of resolutional type are equal to the problems of consistency and completeness of the set of axioms of given theory, the problems of consistency and completeness of intelligent system are equal to the problems of consistency and completeness of the knowledge base of given intelligent system.

3. VERIFICATION PROBLEMS

Let us define the notions of non-redundancy, consistency and completeness and describe the demands to knowledge base that prevent the errors appearance. In accordance with the hierarchic knowledge model the knowledge base contains the elements for data (constants, variables), information (functions, predicates), knowledge (Horn clauses and groups of Horn clauses with common main predicate inside each of the groups) and chains of inference (groups of Horn clauses, composed in accordance with the principle of logical consequence).

3.1. Non-redundancy

3.1.1. Defenitions

An object of data, information or knowledge is called nonredundant, if its removal causes changes of the values and/ or properties of objects which are external with respect to it and/or of a set of object, of which it is an element.
If a removal of an object of data, information or knowledge does not bring about changes in the values and/or properties of the object being external with respect to it, and/or of a set of object, of which it is an element, then such an object of data, information or knowledge is redundant.

3.1.2. Requirements to the knowledge base

Let us list the main requirements to the knowledge base, the observance of which prevents redundancy appearing.

3.1.2.1. Requirements for data

d1. Each fundamental indivisible object of data should be represented by one and only one unique constant; each data population should be compared to one and only one sort $\sigma \in \Sigma$ (if there is a set of sorts Σ instead of one universal sort), and to each sort $\sigma \in \Sigma$ one and only one unary predicate $S^\sigma(x)$ should be compared, expressing that variable x belongs to a sort of objects σ.
d2. For each population A_i, with a sort $\sigma \in \Sigma$ compared to it, there should be no populations $A_{i1},...,A_{in} \subset A_i$ so that $A_{i1},...,A_{in}$ are compared to sorts $\sigma_1,..., \sigma_n \in \Sigma$.

3.1.2.2. Requirements for predicates

i0. Constans and sorts, appearing in predicates, should satisfy the nonredundancy requirements for data. A variable in the scope of the quantifier it is bound by, should be defined as belonging to one and only one sort.

i1. A real object should not be explicitly or implicitly represented in more than one argument of a predicate.

i2. A predicate should not be decomposable into a conjunction or disjunction of several predicates, that is, a predicate should represent an atomic relationship between its arguments.

i3. Predicates should not have redundant arguments. A predicate argument is redundant if its values (and, consequently, the removal of predicate) do not influence the predicate truth or falsity.

3.1.2.3. Requirements for clauses

k0. Constants, variables and predicates, encountered in clauses, must satisfy the listed requirements, preventing redundancy; a variable in the scope of the quantifier it is bound by, should be defined as belonging to one and only one sort, i.e., for each clause C and for each variable x, encountered in the clause C, there should be one and only one unary predicate S (x) should be compared, expressing that variable x belongs to a sort of objects.

(k1-k4). A clause body should not contain redundant predicates. A predicate in a clause body is redundant, if its removal does not affect the values and properties of the given clause under any interpretation of the arguments of this predicate.

Let us list some of the most often variants of predicates' redundancy.

k1. Duplicating premises

A clause contains equivalent predicates as its premises.

Scheme: let there be a clause C1: $A(x,y)$<--$B(x,z),B(x,z),C(z,y)$, predicate $B(x,z)$ is a duplicating predicate; clause C1 should be replaced by a clause C1*: $A(x,y)$<--$B(x,z),C(z,y)$.

k2. Dependent premises

A set of premises of a given clause contains a related subset, that is, a subset of premises which can be derived from the rest of premises (utilizing the available knowledge base).

Note. The clause is a premise of a resolution rule and is replaced with the correspondent resolvent.

Scheme: let there be clauses
C1: $A(x,y)$<--$B(x,y)$ and
C2: $C(x,y)$<--$A(x,y),B(x,y),D(x,y)$;
the utilization of predicate $A(x,y)$ in a clause C2 is redundant;
a clause C2 should be replaced by a clause C2*: $C(x,y)$<--$B(x,y),D(x,y)$.

Note. The last requirement can turn out to be semantically ungrounded.

k3. Subsequent premises

Set of premises of one clause contains as a subset set of premises of another clause with conclusion that is different from the conclusion of the first clause and does not belong to the set of promises of the given clause, i.e., let a<--[B] and c<--[D] be clauses, a, c - conclusions, [B], [D] - bodies of the clauses, {B}, {D} - sets of premises, $c \notin \{B\}$; then, if $\{D\} \subseteq \{B\}$ and clause a<--[E] with $\{E\}=\{c\}U\{F\}$ and $\{F\} \subset \{B\}$ is meaningful, the clause a<--[B] must be replaced with the clause a<--[E].

Note. The clause is a resolvent and is replaced with the premise of the correspondent resolution rule.

Scheme: let there be clauses
C1: $\overline{A(x,y)}$<--B(y,z),C(z,u),D(u,x) and
C2: E(z,x)<--C(z,u),D(u,x)
and C1 can be formulated as C1*: A(x,y)<--B(y,z),E(z,x); then C1 is replaced with C1* (C1 is resolvent of C1* and C2).

k4. Unnecessary premises

Premises are unnecessary if they do not influence values and properties of given clause under any interpretation of its arguments.
- semantic redundancy
    Scheme: let there be clause C1: A(x,y)<--B(y,z),C(z,x),D(x,y), and predicates $B(y,\overline{z})$, $\overline{C(z,x)}$, D(x,y) are interpreted so that predicate D(x,y) does not affect the values and properties of the given clause under any interpretation of its arguments; then a clause C1 should be replaced with a clause C1*: A(x,y)<--B(y,z), C(z,x);
- irrelevant and non-defined predicates
    Scheme: let there be clause C1: A(x,y)<--B(y,z),C(z,x),D(u,v), and predicate D(u,v) assumes a true value under all interpretations considered for the clause C1; then the clause C1 should be replaced with a clause C1*: A (x,y)<--B(y,z),C(z,x).
- contradiction of premises
-- explicit contradiction of premises - a clause contains premises and one of the premises is the explicit negation of another
    Scheme: let there be clause C1: P(x)<--Q(x),A(x),~Q(x), premises containing $\overline{Q(x)}$ are unnecessary and are eliminated from the clause containing them; the clause C1 should be replaced with a clause C1*: P(x)<--A(x);
-- semantic contradiction of premises - a clause contains premises and they assume contradictious (opposing) values in all situations under consideration
    Scheme: let there be clause C1: P(x)<--Q(x),A(x),S(x), Q(x) and S(x) exclude each other and assume contradictious (opposing) values in all situations under consideration; premises Q(x) and S(x) are unnecessary and are eliminated from the clause containing them; the clause C1 should be replaced with a clause C1*: P(x)<--A(x)

Note. It can turn out to be semantically ungrounded to consider contradiction of premises as redundancy; contradiction of premises can turn out to be a consequence of some other types of errors and can be considered as clauses self-inconsistency.

3.1.2.4. Requirements for groups of clauses

gc0. Clauses contained in group of clauses should satisfy the requirements of non-redundancy.

Groups of clauses with a common main predicate

(g1-g7). Group of clauses should not contain redundant clauses. A clause in a group of clauses is considered to be redundant, if its removal does not affect a set of clauses, which can be deduced from the given group of clauses (by resolution).

Let us list some of the most frequently encountered cases of clauses redundancy.

g1. Duplicating clauses
    Clauses contain unified or equivalent premises and conclusions.
    Scheme: let there be clauses
C1: $\overline{A(x)}$<--B(x)
C2: A(a)<--B(a),
clause C2 is redundant as a specific case of clause C1 and should be removed from the group of clauses containing it.

g2. Dependent clauses

A group of clauses contains a related clause (related subset of clauses) if the clause (subset of clauses) can be derived from the rest of clauses of this group (by resolution).

Scheme: let there be clauses
C1: $\overline{A(a,b)}$ <--
C2: A(x,y) <--A(z,y),B(x,z)
C3: A(x,b) <--B(x,a),
clause C3 should be removed from the given group of clauses as dependent clause - it is the resolvent of the clauses C1 and C2.

Note. The generalization of this requirement can be formulated as follows: there should exist one and only one inference for all the clauses that can be derived from the clauses of the given group (by resolution).

g3. Clauses with subsequent premises

Set of premises of one clause contains as a subset set of premises of another clause, the conclusions of the clauses are equivalent.

Scheme: let there be clauses
C1: $\overline{A(x,y)}$ <--B(x,y) and
C2: A(x,y) <--B(x,y),D(y,x)
clause C2 is redundant, because it contains a redundant premise; clause C2 should be removed from the group of clauses containing it.

g4. Clauses with unnecessary premises
- contradiction of premises
-- explicit contradiction of premises - the conclusions of clauses are equivalent, a premise in one clause is the explicit negation of a premise in another clause, the rest of premises are equivalent

Scheme: let there be clauses
C1: $\overline{Q(x)}$ <--B(x),P(x),R(x)
C2: Q(x) <--B(x),~P(x),R(x);
it is naturally to suppose that the premises containing P(x) do not influence values and properties of the given clauses under any interpretation are unnecessary; the premises are eliminated from the clauses containing them; the clauses C1 and C2 should be replaced with a clause C: Q(x) <--B(x),R(x)

Note. Clause C is the resolvent of the clauses C1 and C2;
-- semantic contradiction of premises - the conclusions of clauses are equivalent, there exist premises assuming contradictious (opposing) values in all situations under consideration, the rest of premises are equivalent

Scheme: let there be clauses
C1: $\overline{P(x)}$ <--Q(x),A(x),S(x)
C2: P(x) <--Q(x),B(x),S(x),
A(x) and B(x) assume contradictious (opposing) values in all situations under consideration; premises A(x) and B(x) are unnecessary and are eliminated from the clauses containing them; the clauses C1 and C2 should be replaced with a clause C: P(x) <--Q(x),S(x)

Note. It can turn out to be semantically ungrounded to consider contradiction of premises as redundancy; contradiction of premises can turn out to be a consequence of some other types of errors and can be considered as clauses inconsistency.

g5. Tautologies

Clauses with a (single) premise that is equivalent to the conclusion.

Scheme: clauses of the type C(x,y) <--C(x,y); tautologies are removed from groups of clauses containing them.

g6. There should be impossible to replace a pair of clauses or more clauses of the given group with the less clauses that can be derived from the clauses of the given group by resolution (logical reduction of the group of clauses).
Scheme: let there be clauses
C1: $Q(x)$<--B(x),P(x),R(x)
C2: $Q(x)$<--B(x),~P(x),R(x),
clauses C1 and C2 are replaced with their resolvent C: $Q(x)$<--B(x),R(x); this replacement is justified and necessary.
Note. The possibility of the replacement of certain clauses of the given group with the resolvent of some clauses of this group is determined by the semantics of the participant clauses and may depend upon the context (i.e., clauses of the other groups of clauses).
g7. It should not be possible to decompose a group of clauses into several subsets so that in all possible situations only the clauses of one of the subsets should be used (irrelevant defenition of the group of clauses).
Scheme: let a group of clauses admits the following decomposition:
a<--[B]
a<--[C]
a<--[D],
then the given group can be replaced with the following groups of clauses:
b<--[B]
c<--[C]
d<--[D]
a<--[A],
where [A] determines the situations of the first three groups application.

Chains of inference

c1. Dependent clauses
A group of clauses considered as a chain of inference contains a related clause (related subset of clauses) if the clause (subset of clauses) can be derived from the rest of clauses of this group (by resolution).
Scheme: let there be clauses
C1: $C(x)$<--A(x)
C2: $D(y)$<--C(y)
C3: $D(x)$<--A(x),
clause C3 should be removed from the given group of clauses as dependent clause - it is the resolvent of the clauses C1 and C2.
Note. The generalization of this requirement can be formulated as follows: there should exist one and only one inference for all the clauses that can be derived from the clauses of the given group (by resolution).
Scheme: let there be clauses
C1: $B(x)$<--A(x)
C2: $D(x)$<--B(x)
C3: $C(x)$<--A(x)
C4: $D(x)$<--C(x),
then the clause C: $D(x)$<--A(x) can be derived in two different ways: as the resolvent of clauses C1 and C2 and as the resolvent of clauses C3 and C4; the group of clauses C1, C2, C3, C4 is redundant.
c2. There should be impossible to replace a pair of clauses or more clauses of the given group with the less clauses that can be derived from the clauses of the given group by resolution (logical reduction of the group of clauses).
Note. The possibility of the replacement of certain clauses of the given group with the resolvent of some clauses of this group is determined by the semantics of

the participant clauses and may depend upon the context (i.e., clauses of the other groups of clauses).

Note. The redundant fragments described can be semantically grounded and demand no correction; however, their existence should be revealed and taken into account during the process of inference.

3.2. Consistency

3.2.1. Defenitions

Let us introduce the notions of consistency.

A clause C is consistent if it assumes value "true" at least in one of the interpretations.

A set of clauses S, comprising a knowledge base, is consistent (clauses, comprising a knowledge base, are consistent with each other), if there exists no clause C such that clauses C and ~C can be both derived from the clauses set S (by resolution).

If a set of clauses S, comprising a knowledge base, is consistent, and a set S1=SUC obtained while complementing a consistent clause C to the set S is also consistent, than a set of clauses S and a clause C are consistent with each other (clause C is consistent with a set of clauses S).

For the case of a static knowledge base, as well as with respect to clauses, encountered in knowledge base, a term noncontradictority will also be used; we shall speak of consistency of the existing knowledge base with a new portion of knowledge added to this base.

3.2.2. Requirements to the knowledge base

Formal definitions cover the cases of logical contradictions. Other sorts of anomalies are possible in a knowledge base - cycling, semantic contradictions and etc.

Let us introduce the main requirements, excluding contradictions, for a set of a wider class of clauses compared to Horn clauses and allowing the appearance of negation in premises and conclusion. This consideration is not deprived of interest for practical implementations, permitting the use of negation as a rule, and it conventionally shows possible non-explicitly expressed semantic contradictions.

3.2.2.1. Requirements for clauses

k1. Clauses should not contain contradictions (self-contradictory clauses)

Let us list the main cases of self-contradictory clauses.
- contradiction between the conclusion and premises - one of the premises is the explicit negation of the conclusion

Scheme: clause C: $P(x)<\!\!-Q(x),\!\sim\!\!P(x),\!R(x)$ (C: $\sim\!\!P(x)<\!\!-Q(x),\!P(x),\!R(x))$ is a self-contradictory clause; correction requires an additional analysis;
- contradiction of premises
-- explicit contradiction of premises - a clause contains premises and one of the premises is the explicit negation of another

Scheme: clause C: $P(x)<\!\!-Q(x),\!A(x),\!\sim\!\!Q(x)$ is a self-contradictory clause; correction requires an additional analysis;

-- semantic contradiction of premises - a clause contains premises and they assume contradictious (opposing) values in all situations under consideration

Scheme: let there be clause C: $P(x)\text{<--}Q(x),A(x),S(x)$, $Q(x)$ and $S(x)$ exclude each other and assume contradictious (opposing) values in all situations under consideration; correction requires an additional analysis

Note. Contradiction of premises can be considered as redundancy of the premises.

k2. Clauses should not contain cycles (self-cycling clauses).

A clause is self-cycling clause if it contains a premise equal to the conclusion.

Scheme: clause C: $Q(x,y)\text{<--}Q(x,y),P(x),R(y)$ is a self-cycling clause; correction requires an additional analysis.

3.2.2.2. Requirements for groups of clauses

gc0. Clauses contained in group of clauses should satisfy the requirements of consistency, i.e., no self-contradictory and self-cycling clauses should be contained in group of clauses.

Groups of clauses with a common main predicate

g1. Group of clauses should not contain inconsistent clauses (contradictory clauses).

Let us list the main cases of contradictory clauses.
- pairs of clauses with equivalent conclusions and contradictory premises
-- explicit contradiction - the conclusions of clauses are equivalent, a premise in one clause is the explicit negation of a premise in another clause, the rest of premises are equivalent

Scheme: clauses
C1: $Q(x)\text{<--}B(x),P(x),R(x)$
C2: $Q(x)\text{<--}B(x),\sim P(x),R(x)$
are contradictory clauses; correction requires an additional analysis

Note. Contradiction of clauses can be considered as redundancy of the clauses.
-- semantic contradiction of premises - the conclusions of clauses are equivalent, there exist premises assuming contradictious (opposing) values in all situations under consideration, the rest of premises are equivalent

Scheme: clauses
C1: $P(x)\text{<--}Q(x),A(x),S(x)$
C2: $P(x)\text{<--}Q(x),B(x),S(x)$,
where $A(x)$ and $B(x)$ assume contradictious (opposing) values in all situations under consideration, are contradictory clauses; correction requires an additional analysis

Note. Contradiction of clauses can be considered as redundancy of the clauses.

g2. Group of clauses should not contain clauses with absent satisfactory terminal conditions, i.e., clauses, that lead to cycling. Clauses with the reduction of arguments (argument of the predicate contained in conclusion is a more general case of the argument of the predicate contained in the set of premises or vice versa - argument of the predicate contained in conclusion is the specific case of the argument of the predicate contained in the set of premises) lead to cycling, if the satisfactory definition for specific (general) case of argument is absent in the group of clauses containing the given clause.

Scheme: clause C1: $P(x)\text{<--}P(a)$ leads to cycling if terminal condition for $P(a)$ (for example, $P(a)\text{<--}$) is absent in the group of clauses; correction requires an additional analysis;

clause C2: $P(a)\text{<--}P(x)$ leads to cycling if a clause (group of clauses) defining $P(x)$ is absent in the group of clauses; correction requires an additional analysis.

Chains of inference
c1. Group of clauses should not contain inconsistent clauses (contradictory clauses).

Let us list the main cases of contradictory clauses.
- pairs of clauses with equivalent premises and contradictory conclusions
-- explicit contradiction - clauses with the unified premises, the conclusion of one clause is the explicit negation of the conclusion of another clause
   Scheme: clauses
   C1: $\overline{Q(x)}$<--P(x) and
   C2: ~Q(a)<--P(a)
are contradictory clauses; correction requires an additional analysis;
-- semantic contradiction - the premises of clauses assume equal values in all situations under consideration, the conclusion of one clause is the explicit negation of the conclusion of another clause
   Scheme: clauses
   C1: $\overline{Q(x)}$<--P(x) and
   C2: ~Q(x)<--R(x),
where P(x) and R(x) assume equal values in all situations under consideration, are contradictory clauses; correction requires an additional analysis;
- sets of clauses with the conclusions contradictory to the premises - a premise of one of the clauses is a conclusion (a conclusion of logical consequence) of another clause, the conclusion contradicts a premise of this clause (is a negation of a premise of this clause)
   Scheme: clauses
   C1: $\overline{Q(x)}$<--P(x)
   C2: ~P(x)<--Q(x)
   or
   C1: Q(x)<--P(x)
   C2: R(x)<--Q(x)
   C3: ~P(x)<--R(x)
are contradictory clauses; correction requires an additional analysis
   Note. If the case is that last clause is omitted, but the rule expressed by it is legal in the domain area, implicit contradiction due to semantic gap arises;
- sets of clauses with the contradictory conclusions - a premise of one of the clauses is a conclusion of logical consequence of another clause, the conclusion contradicts the conclusion of this clause (is a negation of the conclusion of this clause)
   Scheme: clauses
   C1: $\overline{Q(x)}$<--P(x)
   C2: R(x)<--P(x)
   C3: ~Q(x)<--R(x)
   or
   C1: W(x)<--P(x)
   C2: Q(x)<--W(x)
   C3: R(x)<--P(x)
   C4: ~Q(x)<--R(x)
are contradictory clauses; correction requires an additional analysis
   Note. If the case is that last clauses of the above sets are omitted, but the rules expressed by them are legal in the domain area, implicit contradiction due to semantic gap arises.
c2. Group of clauses should not contain cycling fragments.

Cycling fragment is a set of clauses, which, being ordered in a logical chain, form a circle.

Scheme: clauses
C1: $Q(x) \leftarrow P(x)$
C2: $R(x) \leftarrow Q(x)$
C3: $P(x) \leftarrow R(x)$
form cycling fragment; correction requires an additional analysis.

Note. If the case is that last clause is omitted, but the rule expressed by it is legal in the domain area, implicit cycling due to semantic gap arises.

3.3. Completeness

3.3.1. Definitions

Let us give two definitions of an intelligent system knowledge base completeness.

A knowledge base is complete if any true statement of the domain area is a logical consequence of knowledge base clauses (can be derived from knowledge base clauses by resolution).

A knowledge base is complete if any logical consequence of knowledge base clauses (a clause that can be derived from knowledge base clauses by resolution) expresses a true statement in the domain area.

There is no sense to consider a problem of completeness as a problem of knowledge base exhaustively describing some domain area, and to require that any sentence true in the domain area could be derived from knowledge base clauses. It is especially important to indicate cases causing incompleteness and to use mechanisms identifying these cases for incompleteness being revealed.

3.3.2. Requirements to the knowledge base

gc1. There should be no situations of missing clauses in a knowledge base. Missing clauses are the clauses which are absent in a knowledge base but are necessary for obtaining some conclusions and the existence of which is real; missing clauses are also mentioned in a case, when arguments, specified in the main predicates of some group of clauses, do not cover all possible argument values.

gc2. A knowledge base should not contain unreachable clauses - clauses with the premises that can never be satisfied, i.e., are not conclusions of other clauses, are not introduced by user or exclude each other.

gc3. (In case of inference from premises to the conclusions)

A knowledge base should not contain terminating clauses.

Terminating clauses are clauses which can be satisfied, but their conclusions are not the goal of a question and are absent among premises of other clauses.

Knowledge base varification is considered to be completed, if the checks for redundancy, inconsistency, incompleteness are carried out, sources of redundancy, inconsistency, incompleteness are localised and necessary correction is implemented.

4. MAIN ERRORS, METHODS OF THEIR DETECTION AND CORRECTION

In conclusion we are going to give a table of main errors, the detection of which is considered to be the most purposefull, accounting for seriousness of errors and the cost of their detection; each type of errors is compared to the method of errors detection and correction.

Table 1

error type	method of detection	correction
redundancy		
1.duplicating premises	a pattern directed search (using unification)	redundant predicates are removed from clauses containing them
2.dependent premises	a search for input inference	correction requires an additional analysis
3.subsequent premises	a search for input inference	correction requires an additional analysis
4.redundant clauses	a search for input inference, search in dependence charts [5]	correction requires an additional analysis
inconsistency		
1.self-contradictory clauses	a pattern directed search (using unification)	correction requires an additional analysis
2.self-cycling clauses	a pattern directed search (using unification)	correction requires an additional analysis
3.contradictory clauses	search in dependence charts	correction requires an additional analysis
4.cycling fragments	search in dependence charts requires	correction an additional analysis

Table 1 (continue)

error type	method of detection	correction
incompleteness		
1. missing clauses	a search for input inference	a hypothesis about the structure of missing clause is put forward
2. unreachable clauses	a pattern directed search (using unification)	correction requires an additional analysis
3. terminating clauses	a pattern directed search (using unification)	correction requires an additional analysis

5. REFERENCES

1. A.F. Kolchin and S.A. Zykova, The intelligent technological design system, Proceedings of joint conference on technological design mechanization in machine-building, Minsk, 1989, 1OO-1O2 (in Russ.).

2. E. Mendelson, Introduction to Mathematical Logic, Moscow, Nauka, 1976 (in Russ.).

3. Ch. Chang and R. Lee, Symbolic Logic and Mechanical Theorem Proving. Academic Press, N.Y., 1973.

4. John.K Debenham, Normal forms of rule-based knowledge systems, Knowledge-based systems, Vol.2, No 3, September, 1989.

5. T. Nguyen, W. Perkins, T. Laffey and V. Pecora, Checking Expert system knowledge bases for consistency and completeness. Proceedings of the 9-th International Joint Conference on Artificial Intelligence, Los Angeles, California, August, 1985, 375-378.

Increased productivity and reliability by intelligent software support

Prof. Dr.sc.techn. D. Kochan, Dr.-Ing. A. Nestler, Dr.-Ing. Chr. Schöne

Abstract

After the characteristic of the state of the art in NC-programming a general frame system and knowledge based methods for the optimization of working values speed and feed will be explained. Further developed modules dealt with the determination of working values for every kind of milling operations.

1. Introduction

The development of the NC-technology in the US-industry was mainly influenced by the lack of qualified workers. In connection with the first industrial application of NC-machine tools therefore NC-programming systems for external support of the engineers were developed.

The primary aspect of this first generation of NC-programming language (APT-automatically programmed tools) and its derivades IFAPT - France; NEL-APT - United Kingdom; FAPT - Japan) was mainly the determination of gemetrical information in connection with the tool-path generation

Starting with the NC-Programming system EXAPT (Extension of APT), SYMAP and other ones /1/ more and more technological support was included into the NC-programming systems.

Nowadays two important trends can be pointed out:

1. Integration of all essential decision making in connection with geometrical, technological, economic and organisational problem solving in the entire process chain (design, calculation, process and operational planning, process implementation).

2. Increased application of human knowledge and experiences of the programming systems.

Based on international developing trends and our own results some important experiences will be demonstrated.

2. State of the art: NC-programming and possibilities for user support

Current development trends in the entire CAP-field are mainly concentrated on the interface of human and CAP-system. The NC-programming in the framework of CIM-environments is mainly influenced by CAD/NC-coupling, integration and realization of process feedbacks.

Regarding to the broad industrial application data transfer starting from CAD into NC-programming systems is in a starting phase. Most of the applied solutions based on coupling methods are realized by kind of standard interfaces and restricted to 2D-geometries. Additionally to this development stage the availability of integrated systems based on CAD-systems is more and more increasing. Such solutions are characterized by a full CAD-functionality for NC-programming, but they are not capable to fulfil in the same mode the necessary CAP-functionality of powerful traditional NC-programming systems.

Based on statistical analyses it will be estimated that currently data transfer is realized as follows /2/:

50 % via drawing
30 % via standard interfaces
10 % via macroparameter and
10 % integrated models

The current situation of data-handling at powerful NC-programming systems can be characterized from the point of view of the user by the following features:

- taking over CAD-geometry via standard-interfaces,
- unified programming language for all machining procedures,
- administration (supervising) of NC-machining macros in macrolibraries,
- determination of technology data on the basis of tool and cutting value files,
- tool databank for tool management,
- NC-archives for the management of NC-control information,
- machining simulation and
- postprocessor generation.

For the NC-data generation are one way interfaces as they are called from the point of view of information technology are used /3/:

- product definition information (IGES; DXF;...)
- process describing information (ISO 4342)
- machine tool neutral information (ISO 4343, ISO 3592)
- machine tool related information (ISO 6983)

Additionally to geometry taking-over the real task of the NC-consists in programmer including manifold process describing information for the generation of movement orders.

Currently all powerful NC-programming systems are capable to solve all kinds of programming tasks for all manufacturing technologies within 5-axis machining. An important quality criterion for technological planning of sequences is the support at different automatized functions, such as:

- determination of cutting values via material and tool specific tables or optimization models for cutting values,
- subdivision of cuts at turning,
- optimization of milling path for pocket machining or
- optimization of the operational sequence with tool selection for boring.

Additionally to the unified CAP-functionality by automatized sequences unified user-interfaces will also be available. But only with user comfort and help functions operational planning cannot reach a new quality level. The same comment can be given for acceleration of programming by data bank access and using parameterised macros.

In general the broad application of specific technology knowledge is underdeveloped. Additionally to the so called "classic requirements" for efficient possiblities of such application systems more and more

- required preprocessed technology knowledge and
- effective process data simulation

will be required.

Especially for complex manufacturing problems some specific experiences are necessary for defined combinations of equipment in connection with determined materials and determined cutting values in a determined operational-regime. In a real factory environment such knowledge is available in the form of tables, instructions, experiences or etc.. This must be disposed clearly, unmistakably and in a simple way.

The current performance - reached by conventional programming techniques additionally to optimizing and simulation strategies to be completed by knowledge based methods. The aim is to be characterized by the following steps:

- to open knowledge sources
- to refine knowledge and
- to make available stored knowledge with new solutions (fig. 2.1.) /4/

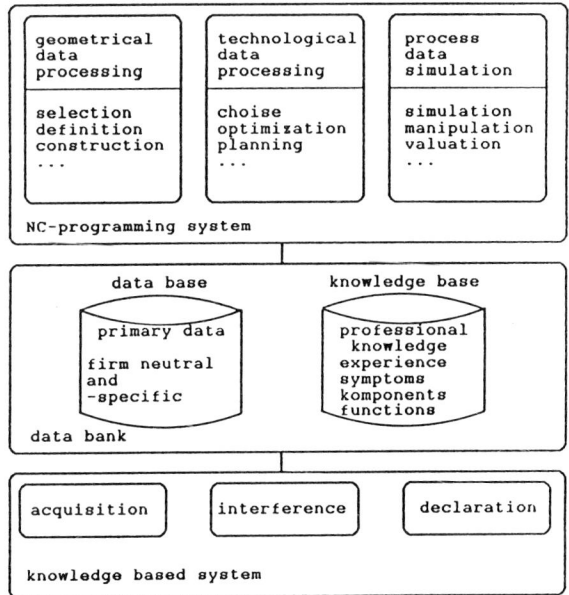

Fig. 2.1.: Komponents of a knowledge based approach in the NC-programming

Most important in this connection is the socalled "implicite experience-knowledge" /8/. This category means the knowledge which results from active handling of sophisticated operations in an automatized manufacturing environment.

3. Knowledge based support for the NC-programming

The first essential step is knowledge acquisition of the technological state of affairs /5/. The operational planner and NC-programmist will be supported in making available and structuring the necessary information. In this way step by step a knowledge based background can be created for operational planning and NC-programming by different sources:

- objective general knowledge and
- subjective specific knowledge of the user.

Selected aims can be directed to

- gaps in the information flow
- quality and reliability of planning results or
- extension of data base.

This knowledge based background can include problem areas such as:

- functional technological relations
- criteria for applications
- guiding principles for applications or
- factory specific know-how.

It is more difficult to get a personal specific know-how. Even for mathematical models the prepared knowledge could be a necessary prerequisite for supporting data handling. Concerning CIM-requirements the autonomy and suboptimality of the optimization results of cutting conditions has to be dismantled step by step. First steps can be realized by

- integration of different manufacturing procedures (e.g. turning, milling) also by

- embedding into a knowledge based background (see fig. 3.1.).

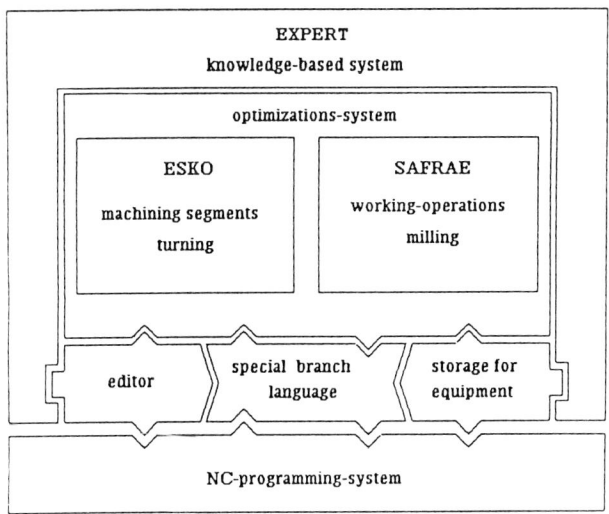

Fig. 3.1.: Embedding of optimization of working values into a knowledge based NC-programming

Integrated solutions are necessary, which combine conventional and knowledge based parts. An obvious example is the integration of possibilities of knowledge based manipulation of the optimization of working values in the framework of NC-programming /6/. The integration with a knowledge based system component leads to the effect that the multiplicity of influence parameters of the technical constraints will be surveyable even for not so richly experienced users. Only in this way the complexity of the sophisticated matter can be mastered and handled (see fig. 3.2.). Especially for planning variants and the application of different equipment the knowledge of the relationships between different influence factors is indis- pensable. By this intelligent upgrading it was contributed essentially to the user acceptance of an optimization model.

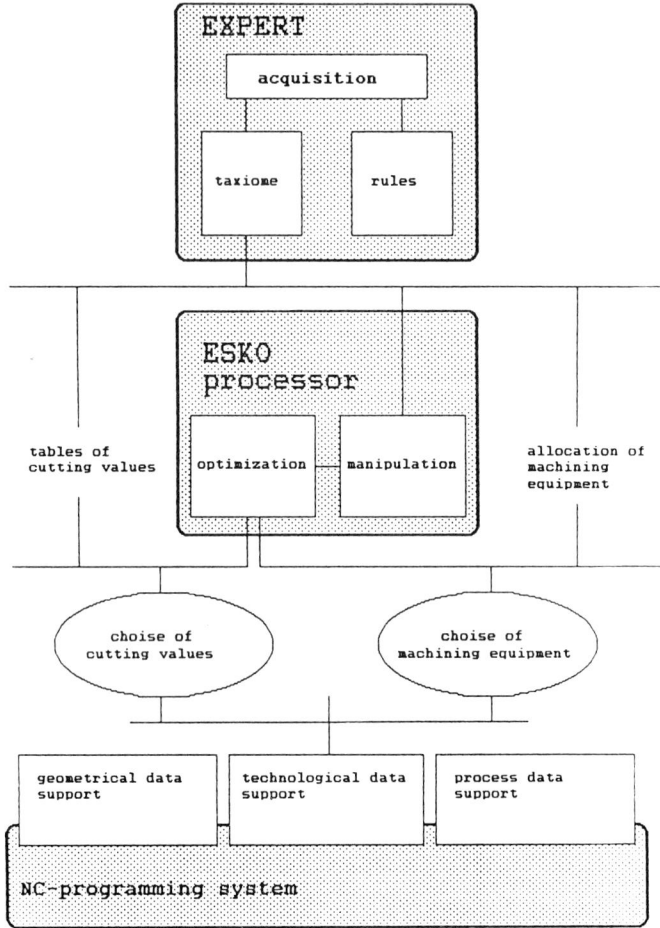

Fig. 3.2.: Alternatives for technological data support with integration of knowledge based methods

The CIM-oriented NC-programming environment requires a comprehensive information management additionally to the necessary knowledge basis. The necessary information systems perform especially the supervision of manufacturing oriented information and drawings. The kernel of such systems is nowadays the data bank supported, flexibly usable data management, normally with the help of index registers or by direct access.

In this relation the development of data banks with manufacturing data is of high importance. They are essential prerequisites for the economy of computer aided workplaces and also for the integration of automatization islands. The NC-programmer and operational planner will be provided a manufacturing data bank with a user accepted working environment. This will allow him quick access to stored sequences, rules, data and documents.

In cooperation with the Research Association Programming Languages e.V. Aachen a manufacturing data-bank will be developed for NC-programming and operational planning considering particularly the requirements of small and mediumsized companies. In the framework of this research project the manufacturing data for

- equipment and methods,
- cutting values and normatives,
- user elements (feature, macro),
- manufacturing technology functions,
- adaption to the NC-machine tools and
- manufacturing-data

will be taken into account (see fig. 3.3.)

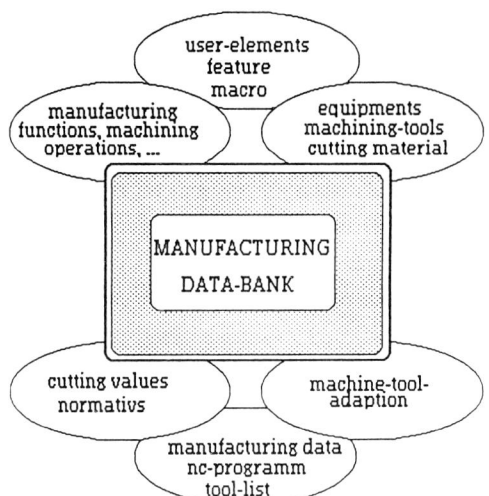

Fig. 3.3.: Main komplexes of the manufacturing data-bank

The following aims are in the foreground of activities:

- interchangeability of the manufacturing data stock;
- balancing of different data, methods and procedures;
- consideration of standard interfaces and
- consistent know-how utilization.

The conception and realization of such a project required a systematic approach in the preparation phase /7/:

- determination of aims and nomination of a requirement list;

- function and information analyses in selected enterprices in connection with a data-analysis as a basis for the data model, which has to be developed.

With the developing trends of knowledge processing and manufacturing data bank towards for operational planning and NC-programming it will be contributed to intelligent software support.

4. Determination of working values for every kind of milling operations

The currently available CAD/CAM-solutions for free-form shapes are characterized by a high level of the geometrical components. On the other hand, it can be pointed out that the support for the operational planning is very low. For decision support from the manufacturing point of view there are mainly tables of the cutting-values available. Such in the form of tables are available of external tables (conventional way) or internal tables in data files as part of the CAD/CAM-system. The entire manufacturing system with to main components:

- tool
- chuck
- machine tool
- cutting material
- workpiece

can't be considered in its total complexity. Usually decision making concerning tool selection and milling path determination is the task of the NC-programmer. However, due to the restricted knowledge in connection with the above emphasized high complexity of the problem field there is an objective gap given. In many cases are the results:

- failure concerning the manufacturing process
- economic disadvantages
- low quality results, which required additional manual operations

For avoiding these mistakes and disadvantages a more user-friendly solution of our module for the optimized determination of working values SAFRAE /9/ was developed. This step into direction of increased technological performance was realized by integration into the CAD/CAM-system TECHNOVISION - Norsk Data.

Based on the modelling of essential restrictions of the cutting process at the programm complex SAFRAE important technical aids for the technological and economical decision making are available (fig. 4.1.). The used models reflect the sum of technological knowledge. Obviously they are influenced by continious dynamic change. Most important for all calculations are in this context quality, volume and the to date available primary data of tools, adapter, machine tools and cutting materials.

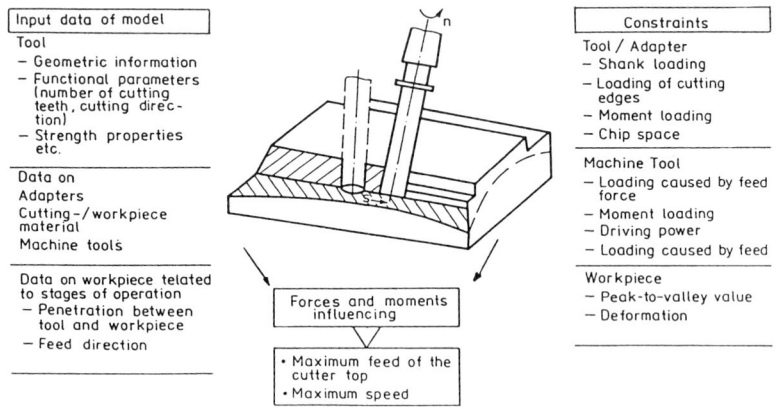

Fig. 4.1. Technological model of 3D-5D Milling Operation described by input data and constraints

Furthermore, the acceptance of cutting modules with high performance depends on availability and handling support. As part of the CAD/CAM-system TECHNOVISION the module TECHMILL is developed for the generation of NC-control data for workpieces with free-form shapes. For the rough machining operation a coupling was realized (see fig. 4.2.). The main influence factors for rough machining in connection of workpieces with free-form shapes are additionally to the determination of technologically and economically profound working-values especially the

- suitable selection of tools,
- determination of milling path and
- overlapping of milling path.

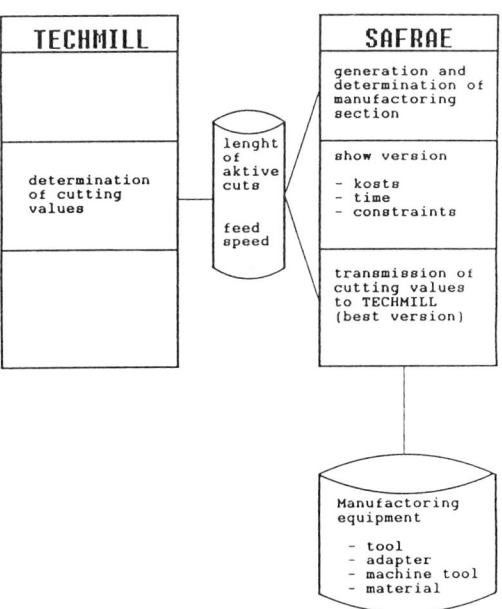

Fig. 4.2.: Connection TECHMILL and SAFRAE

In this cases the decisions are influenced interchangeably. The overlapping of the milling path is a relative value of the radial cutting depth (equation 1)

$$a_{rrel} = (d - a_r) / d$$

a_{rrel} : *relative overlapping of tools*

a_r : *radial cutting-depth in mm*

d : *tool diameter in mm*

If for one machining variant the tool diameter is fixed so for a determined milling regime the variable values are

- cutting values
- cutting depths and
- radial depths of cutting

This leads to a 4-dimensional optimization problem (equation 2).

$$t = f(v_f, n, a, a_r) = l\,(a, a_r) / (\,V_f\,(a, a_r) * n\,(a, a_r)\,)\ \text{MIN}$$

t : lead time (?)   in min
V_f : feed   in mm
n : speed   in 1/min
l : tool path   in mm
a : cutting depths   in mm
a_r : radial cutting depths in mm

The restrictions for the cutting depths and radial cutting-depths based on user experiences as it is demonstrated in fig. 4.3.

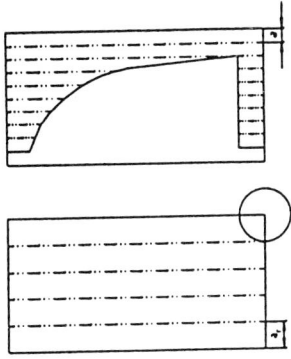

Version	d cutter diameter /mm/	a depth of cutting /mm/	a_r width of cutting /mm/	l lenght of active cuts /mm/	t time of cut /min/
1	10	5	10	57941	603
2	10	3	8	103933	503
3	16	8	12	38021	87
4	16	8	8	40161	71

Fig. 4.3.: Versions of rough machining

Because currently an explicite calculation of

- optimized tool diameter,
- the milling path-overlapping and
- leveling

is not possible we oriented to a variant approach. That means the 4-dimensional problem will be reduced to a 2-dimensional one. At first with the aid of TECHMILL variant of rough-machining will be generated. As a result the milling-path and the length of the active cuts will be calculated cuts.

After the activation of program SAFRAE at first the input of the equipment in the form of code numbers has to be done, then the cutting specific data follow: cutting depths, radial cutting depths and the lengths of the active tool pathes. For every machining variant the cutting-values the main times (lead-time) and the machining costs will be calculated. The loading of the single technical constraints per cut are clearly demonstrated.

By comparision of the calculated variants concerning cost and time a suitable machining variant can be determined. After that the calculated cutting values will be delivered to the program module TECHMILL for the generation of NC-control information. Based on the used example it can be demonstrated that there is a broad variety of main-times given in connection with the calculation of restrictions with variants of tools, milling path levelling and milling path overlapping (fig. 4.4.).

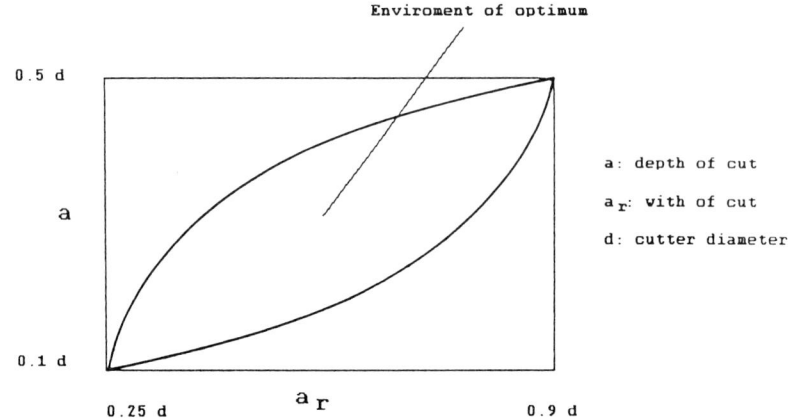

Fig. 4.4.: Constraints of with and depth of out

5. Summary and conclusion

Finally it can be pointed out that the creative engineer is primarily responsible for reliabe and productive manufacturing processes.

But nowadays additionally to his specific experiences he can use a lot of specific software for decision support in connection with powerful CAD/CAM-systems. The results demonstrated possibilities to solve manufacturing problems of high complexity in connection with complicated workpieces.

Literature

/1/ Leslie, W.: Programming Languages for Machine Tools; Rom 1969, PROLAMAT, North Holland Publishing Company, Amsterdam-London

/2/ NC-Programmiersysteme im Vergleich; KfK-CAD/CAM-Karlsruhe, 1990

/3/ Eversheim, W.: CAD-Systeme und NC-Programmiersysteme koppeln; ZWF-CIM, 85, 1990/5, S. 267-271
u.a.

/4/ Pritschow, G.: Künstliche Intelligenz in der Fertigungstechnik; München, Carl Hanser Verlag München, Wien, 1989
Spur, G.,
Weck, M.

/5/ Nestler, A.: Computergestützte Wissensakquisition - Nutzung für technisch-technologische Sachverhalte; ZWF-CIM, 85, 11/1990, S. 580-583
Sämisch, J.

/6/ Nestler, A.: Wissensbasierte Kinematik- und Schnittwertermittlung beim Drehen; ZWF-CIM 84, 1989/6, S. 321-326
u.a.

/7/ Dittmer, H.: Relationale Datenbanken - Kern der rechnerintegrierten Produktion; CIM-Management 6/1989, S. 34 - 39
Lamplemeyer, U.

/8/ Martin, H. :   Computergestützte erfahrungsgeleitete Ar-
    Rose, H.   beit bei Werkstattprogrammierung, Perspek-
                 tiven für Programmierverfahren und Steue-
                 rungstechniken; In Rose, H. (Herausgeber)
                 Programmieren in der Werkstatt, Campus,
                 1990

/9/ Kochan, D.: Komplexe Schnittwertbestimmung für beliebi-
    Schöne, Ch. ge Fräsaufgaben; Zeitschrift für wirt-
                 schaftliche Fertigung, 3/89

MONITORING AND MANUFACTURING ASPECTS

CHAIRMAN: A. STORR
UNIVERSITY OF STUTTGART, GERMANY

Model based Diagnostics of Machine Tools and the Turning Process

Prof. Dr.-Ing. H. Schulz
Dipl.-Ing. H. Schönherr
Dipl.-Ing. K.P. Gebauer

Institute for Production Engineering and Machine Tools (PTW)
Technical University Darmstadt

Abstract
In the following paper a dynamic model for a machine tool and the turning process will be developed by methods of the system-theory principle. The model describes the state at the lip of tool and the machine tool. The parameter estimation method is used to determine the parameters of the mathematic model which is described by differential equations. Between the errors and the corresponding model parameters there is a close correlation. Therefore, failures can be exactly predicted, and fault location and course of failure can be reliably diagnosed.
Examples for the detection of possible failures as tool wear, tool break, chip break, chip lamination and unbalanced workpiece will be shown.
These different process events will be acquired from a tool-monitoring-system realized by PTW, and they will be graphically imparted to the machine-user in an easily understandable form.

1. INTRODUCTION

In order to increase the productivity of the entire plant the complexity of production units and the automatisation of production processes continuously increased during the last years.
The high degree of complexity makes it impossible for the machine-user to see, to understand and to influence the processes of production in the encapsulated machine.
At the same time a development started to manage the production plants with lowest man-power. These developments will be successful if it is possible to acquire the regularities of the cutting process by a monitoring-system, in order to make them transparent for the operator.
In the past the modelling had a phenomenological character. Experimental studies at turning machines have shown that most standstill times of machine

tools result from the tool. The production process and the tool represent the main focus of monitoring.

2. CHIP FORMATION AND IDENTIFICATION

The origination of chip laminae can be considered as a dynamical process with periodical changes of phases of upset, separation and glide. During the process of upsetting the dynamical power amplitude increases, during the process of sliding it decreases.
The chip-laminae loosen with a frequency which is generally named lamination-frequency.
The determination and the command of the chip forms which can be classified after [11], is very interesting.
The chip break frequency, a central factor shows the number of the arising single chips per second.
The chip break generates additional dynamical portions in the power signal which become noticable in the frequency spectrum analoque to the chip lamination frequency (fig. 1).

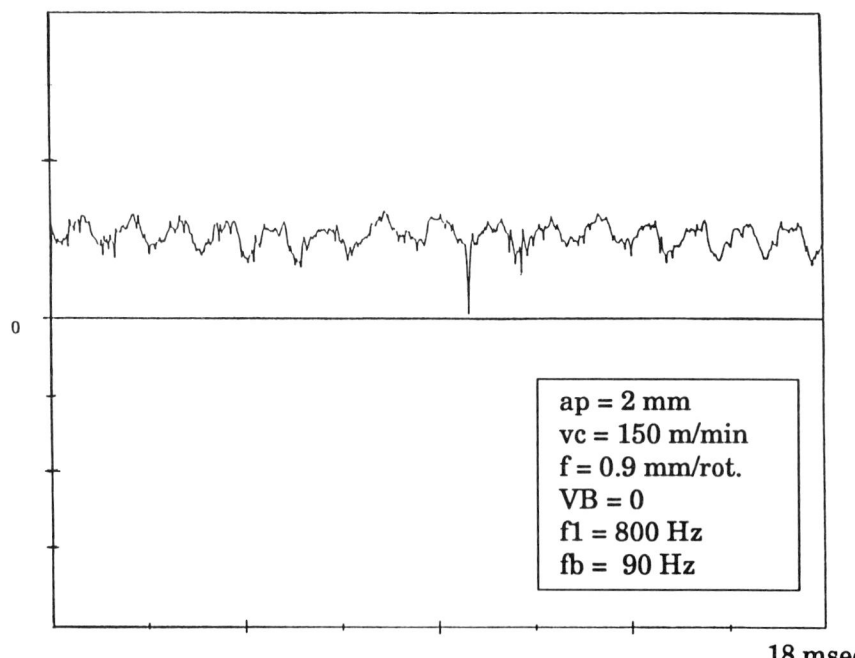

Fig. 1: Feed force

The variation of the chip break frequency and lamination frequency, depending on wear, is shown in fig. 1. From the spectral analysis of the power signal the following information about the process can be determined:
- chip lamination frequency
- chip break frequency
- jammed chips
- unbalanced workpiece
- wear of the tool
- tool grind-in process.

Experience shows that new tool-lips yield bad workpiece surfaces, ground in tool-lips show good results. This is plainly recognizable in the frequency spectrum. A new tool-lip generates a diffuse spectrum in which all frequencies are stimulated with a very high amplitude level. With increasing grind-in duration the amplitude level decreases, while the frequencies which describe the process e.g. lamination, chip break, unbalanced workpiece endure. On this way the grind-in process is controllable and comprehensible.

Another possibility to describe the process events is to consider the cutting process as a scaled, stochastic signal process. This can be described by means of a stochastic differential equation as a parametric ARMA (autoregressive moving average)- model, here shown as a differential equation
with
$V(k)$ = white noise
$Y(k)$ = output signal of an imagined filter

$$Y_{(k)} + C_1*Y_{(k-1)} + C_2*Y_{(k-2)} + \ldots + C_n*Y_{(k-n)} =$$
$$d_0*V_{(k)} + d_1*V_{(k-1)} + d_2*V_{(k-2)} + \ldots + d_m*V_{(k-m)} \tag{1}$$

The entire system behaviour can be described with the ARMA-model, especially the process events: chip lamination, chip breaking, unbalanced workpiece and vibration.

3. IDENTIFICATION OF DYNAMIC SYSTEMS

To describe the behaviour or to acquire the internal regularities of an object, system or a process, the modelling has qualified as an important part. The system theory makes it possible to describe the system behaviour over time with support of mathematic models [1,2,6,7,8,18]. A mathematic model of an object or system represents an operator F, which takes over the transformation of the input variable $X_e(t)$ into the output variable $X_a(t)$.

$$X_a(t) = F[X_e(t)] \tag{2}$$

The operator F describes the entity of all the logical and mathematical operations (addition, multiplication, integration, differenciation etc.), which show a dependency between the functions $X_a(t)$ and $X_e(t)$. It connects the cause i.e. the independent input variable $X_e(t)$ with the effect, i.e. with the dependent output variable $X_a(t)$.

The determination of the dynamical behaviour, i.e. of the system, can be done in 2 ways [6,7,8]:
- theoretical and
- experimental.

With the theoretical modelling the mathematic model will be determined. Based on physical rules e.g. with mechanical systems power and momentum balance, conservation laws for energy, impulse and mass, etc. will be formulated. The theoretical model is generally very complex.

With the experimental modelling the mathematical model will be determined from measurements of the system input and output variables by means of applicable identification processes [6,18]. The system analysis should always be used, supplementing theoretical and experimental processes.

4. MODELLING OF THE MACHINE TOOL

An exact theoretical system analysis of the machine tool results in differential equations of a high degree. The experimental system analysis brings a reduction of degree with an acceptable loss of accuracy and information [5,6].

The proceeding and the correlation invest made, will be shown taking the feed drive of an cnc-lathe (z-axis) as an example.

In order to achieve a transferability onto a wide ranged machine spectrum a modular splitting of the assemblies is necessary (fig.2).

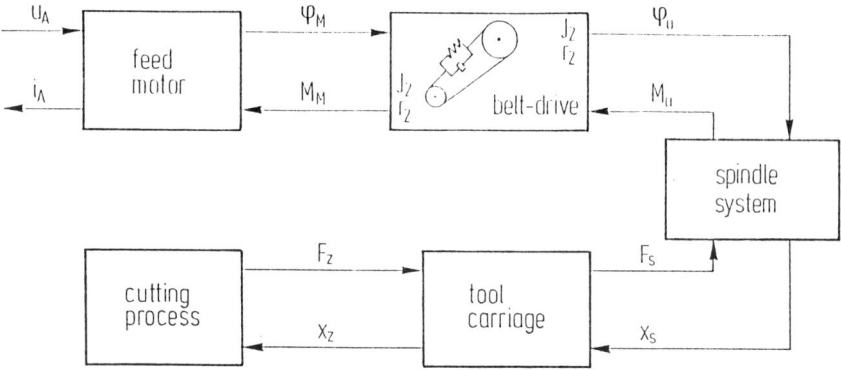

Fig. 2: Model of the feed-drive system

The single part models describe motor, toothed belt drive and feed drive on the base of the regarded theoretical system. The following differential equation is valid for the subsystem belt drive:

$$-M_M = J_1 * \ddot{\varphi}_M + C^*_R * (1/\ddot{u} * \varphi_M - \varphi_G) + K^*_R * (1/\ddot{u} * \dot{\varphi}_M - \dot{\varphi}_G) \quad (3)$$

$$-M_G = J_2 * \ddot{\varphi}_G + C*R*(\varphi_M - \ddot{u}*\varphi_G) + K^*_R * (\dot{\varphi}_M - \ddot{u}*\dot{\varphi}_G) \quad (4)$$

with

$$K^*_R = K_R * r_1 * r_2 \quad (5)$$

$$C^*_R = C_R * r_1 * r_2 \quad (6)$$

To investigate the reference parameters of a new machine tool the closed loop of the feed drive is splitted and the speed set value is externally determined.
As input a noise signal which is limited to the relevant frequency range is used.
A spectrum analyser for measurements' acquisistion calculates the following parameters out of the scanned input and output variables: frequency spectra, amplitude response, phase response and coherence function.
The course of the amplitude spectrum at the reference test of the infeed drive in z-direction is shown in a Bode-diagramme with the speed set value as input value and the actual speed value as output value (fig.3):

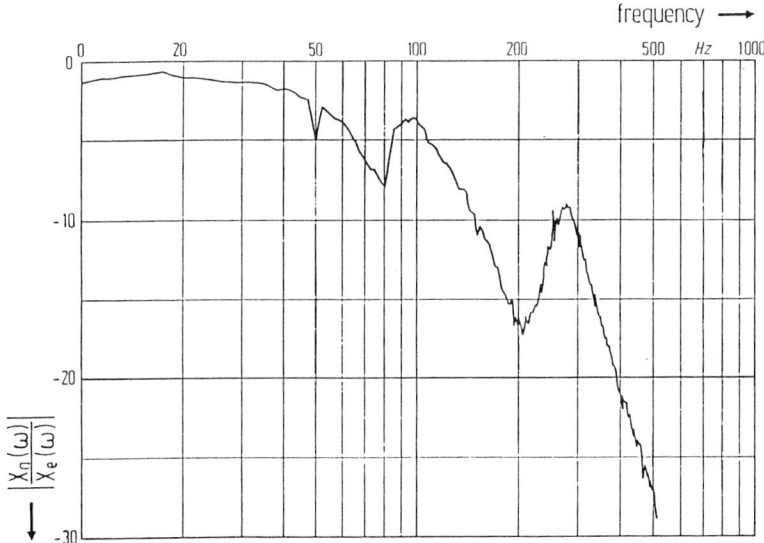

Fig. 3: Reference Spectrum

The diagram contains the following signigicancies:
- 0<f<30 Hz: a rather constant course
- 30>f>70 Hz : dropping with 20 dB/decade
- 85 Hz: resonance position of the feed spindle system
- 270 Hz: resonance position of the toothed belt drive.

This characteristic amplitude course allows the reduction of the entire feed drive system to a model of the fifth degree.

By modification of the system (toothed belt, speed controller, spindle etc.) one can show that the significant spectral ranges and therefore the model parameters can be directly allocated to the specific assemblies. Fig.4 exemplarily shows the changes of the amplitude spectrum with decreasing preloading of the toothed belt. One can clearly see that in this case there only are changes in the second resonance position (at about 270 Hz), while the other ranges of the spectrum are not touched.

This experimentally achieved knowledge concerning the electromechanical system "feed drive z-axis" with the speeds as input and output variables shows that the state of the essential diagnosis relevant components can be exactly diagnosed. Hence, with the temporal changes of the frequency course, a tendency analysis and therefore a forecasting system diagnosis can be done (fig. 4).

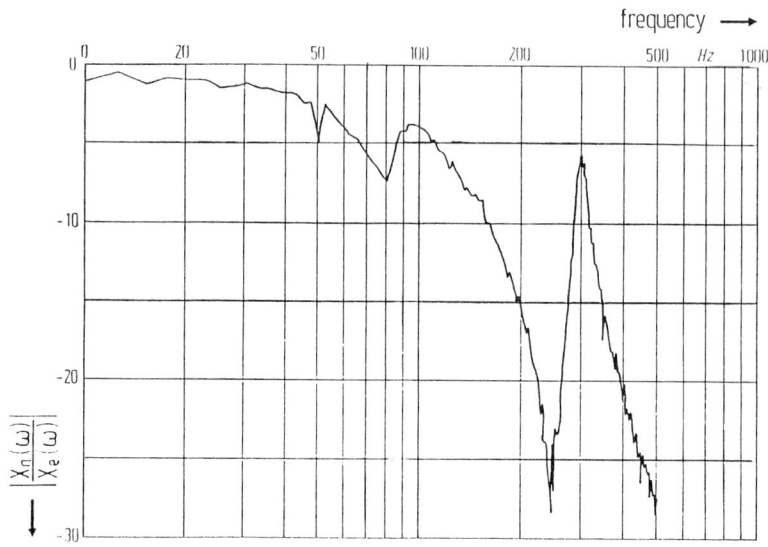

Fig. 4: Spectrum with charge of belt preloading

5. MODELLING OF THE CUTTING PROCESS

5.1 Theoretical Modelling

The basic idea for the conception of the dynamical process-model is the projection of different cutting conditions and variations at the tool lip (wear, micro cracks etc.) in different process parameters [15].

The cutting process is the link between workpiece and tool and builds a complex, threedimensional-effect mechanism. The dynamic process model is base on the fact that the changes at the cutting edge (wear, minor tweak outs etc.) and the cutting conditions are to be seen in the various process parameters.

The cutting process contains the events which take place in the contact zone and which can be described by a spring-damping-system. The material behaviour has to be regarded as it is dominant in the process. To register the kinematics of the entire system one has to notice the physical characteristics of the tool and the workpiece.

Fig. 5 shows the model which is reduced to the linear one-dimensional spring-damper-mass-element-system. For the machining process "longitudinal turning" there is the following movement equation without considering the masses:

$$(K_1+K_2)*q_1 - K_2*\dot{q}_2 + C_1*q_1 = 0 \tag{7}$$

$$-K_2*\dot{q}_1 + (K_2+K_3)*\dot{q}_2 + C_3*q_2 = K_3*\dot{q}_3 + C_3*q_3 \tag{8}$$

The cutting force F is measured on the position 2 in Fig. 5.

$$F = K_2*(\dot{q}_2-\dot{q}_1) \tag{9}$$

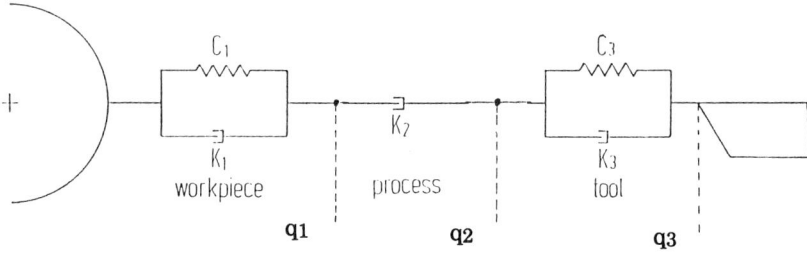

Fig. 5: Mechanical model of a turning process

With the infeed speed q3 and the cutting force F one gets as result from a Laplace transformation the following transfer function:

$$\frac{F_{(s)}}{q_{3(s)}*s} = \frac{K_2*[C_3*C_1+s(C_3*K_1+C_1*K_3) + s^2 (K_3*K_1)]}{C_3*C_1+s(C_1*K_2+C_1*K_3+C_3*K_1+C_3*K_2)+s^2(K_1K_2+K_1K_3+K_2K_3)} \quad (10)$$

Considering the geometric correlations at the cutting into the workpiece which are shown by the transfer function

$$Ga = 1/(1+T_a*s)$$

one gets as a total transfer function a model of the 3rd degree.

5.2 Experimental Modelling

In a second step the determined theoretical model will be experimentally verified. Both, the order as well as the parameters of the model must be determined.

Experiments have shown that a reduction is possible.

Many model parameters include the parameter K_2 which registers wear dependend changes in the process behaviour.

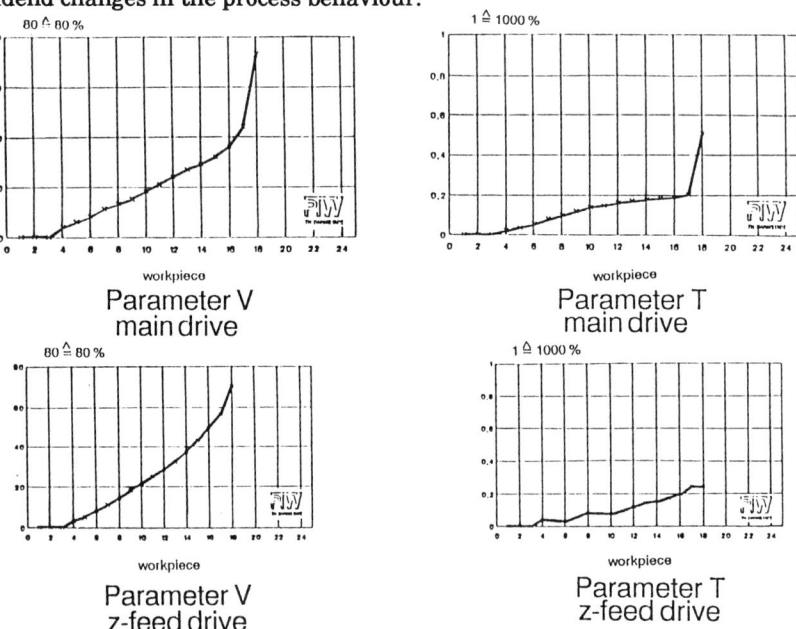

Fig. 6: Relative changes of the parameter V and T in dependence of the workpiece (tool wear)

The estimation of the parameters has been performed with the least square method [6], which has proven true in production engineering. Fig. 6 shows the course of the estimated parameters T and V of the feed- and main drive. At both drives the time constant shows the biggest tool wear dependend changes. If there are micro tweak outs, the time-constant T exponentially increases.
These model-technical cognitions confirm the generally known fact that worn tools are less effective than sharp ones because they need more time to get into the workpiece. Hence, there is a higher time constant. Therefore, this feed drive model can be used by suitable evaluation methods for precise forecasting diagnosis of the tool wear development of the used tool.
The model and the controlling system which is described in 6. was already tested in practice. Here, approx. 1400 workpieces have been produced which had been machined by 6 tools. One got a high hitting accuracy and liability because the diagnosis has been done by changing several model parameters. Therefore the acceptance of the controlling system by the operator was given.
Investigations in our own laboratory showed that even with little potential ranges on the basis of the force models there are many wear dependent changes of parameters which are used for the tool diagnosis.

6. DIAGNOSIS AND FORECAST OF TOOL WEAR

With the conception of the tool monitoring system there always is to distinguish between tool wear diagnosis, and break or collision monitoring, because both cannot be included within the same strategy.
The wear diagnosis concept bases on the idea that the machine user exactly knows as a basis of his specific knowledge which NC-program sentences are critical.
The monitoring system enables the operator to analyse these critical NC-program sentences by deepening the diagnosis.
The message "NC-start" puts the diagnosis computer into stand-by conditions. By reading the cutting number, the measuring of speeds, voltage and currents is triggered (fig.7). The readings are digitally taken, pre-processed and afterwards given to a parameter-identification algorithm whose part referring parameter sets are used for the diagnosis of the tool's state. The cutting number referring parameterisation is done by a controlling file which is located in the diagnosis computer. There, it is determined which of the special single part models (e.g. cutting force, feed force, passive force, feed drive (FD) Z- and X-axis or main drive (MD)) must be identified as a relevant basis for the parameter estimation. The parameter sets which have been estimated for the single part models are used for diagnosis and forecast. Diagnosis and forecast are transparently given to the machine user by a graphic interface.
Because of the three-sectioning "human being-machine-diagnose computer" a sozio-technical system is build whose three components actively communicate.
The tool-break-monitoring effects from the whole processing time of the

workpiece, the measured values will be continuously acquired with a high scanning rate and evaluated by a quick and easy strategy.

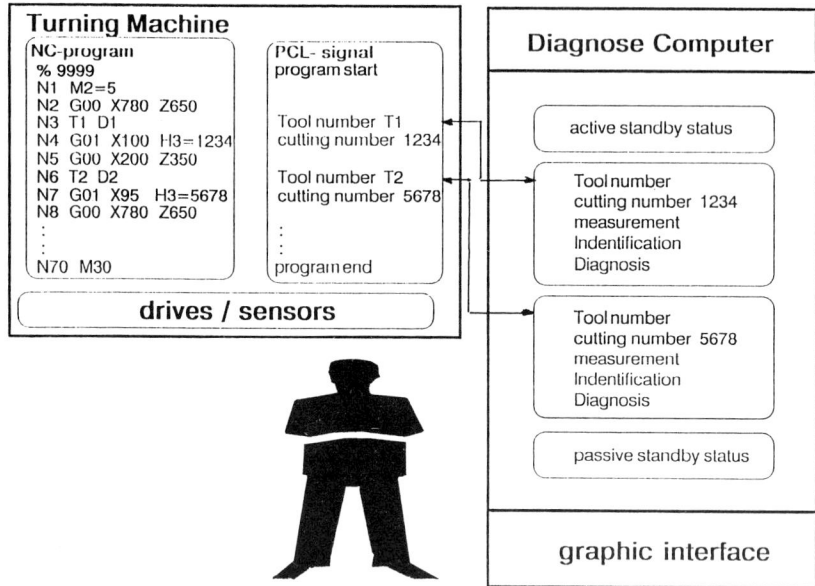

Fig. 7: Tool monitoring system

7. CONCLUSION

The dynamic courses that occur during turning can be described by a PDT2 -transfer function. Significant tool wear dependent changes of the parameters amplification and time constant were described. The parameter time-constant T shows the highest tool wear dependent changes.
For a sensitive and exact tool diagnosis these parameter changes are used. As demonstrated with the example of a feed drive of a cnc-lathe the method to analyse the state of electromechanical components of a machine tool shows the possibility of a secure forecasting diagnosis. The maintenance can be reduced if a periodical maintenance can be done instead of a planned, stress dependent maintenance.
In order to control and diagnose the operating process as well as the machine itself, only already existing control-internal parameters are used. The integrated controlling system is transferrable onto the machine as well as to the cutting process.
The productivity of the entire production unit increases because machine down-times are reduced and major detected failures are avoided.

8. REFERENCES

1. Aström, K., 1968: Lectures on the identification problem, The Least Squares. Report 6808
2. Balahrishna, AV., Peterka, V., 1969: Identification in automatic control system.
3. Dewhurst, P., Collins, I.A., 1973: A matrix technique for constructing slip-line field solutions to a class of plane strain plasticity problems, Int.J.Num.Math.Eng. 7 357-378.
4. Eyckhoff, P., 1974: System Identification, London.
5. Gebauer, K.P., 1990: Fehlerfrüherkennung an elektromechanischen Komponenten moderner CNC-Werkzeugmaschinen, Proceedings "Fehlerfrüherkennung in der spanenden Fertigung", Düsseldorf.
6. Isermann, R., 1988: Identifikation dynamischer Systeme Band 1 und 2.
7. Isermann, R., 1984,: Process fault detection on modelling and estimation methods - a survey, Automatica 20, 387-404.
8. Isermann, R., 1980: Methoden zur Fehlererkennung für die Überwachung technischer Systeme, VDI-Bericht 364.
9. Kluft, W., 1983: Werkzeug-Überwachungssysteme für die Drehbearbeitung, Diss. TH-Aachen.
10. Merchant, M.E., 1945: Mechanics of Metal Cutting Process; J. of Applied Physics, Vol. 16, 267 ff
11. N.N., 1969: Zerspanversuche, Spanbeurteilung, Stahl-Eisen-Prüfblatt 1178-69, Düsseldorf.
12. Piispanen, V., 1948: Theory of Formation of Metal Cutting, Int.J. of Applied Physisc 19, 876 ff.
13. Peschel, M., 1978: Modellierung für Signale und Systeme, VEB Verlag Technik.
14. Schulz, H., 1990: Fehlerfrüherkennung an Werkzeugmaschinen bedeutet Produktivitätssteigerung und Kostensenkung, Proceedings "Fehlerfrüherkennung in der spanenden Fertigung", Düsseldorf.
15. Schönherr, H., 1990: Modellgestützte Fehlerfrüherkennung beim Zerspanprozeß Drehen, Proceedings "Fehlerfrüherkennung in der spanenden Fertigung", Düsseldorf.
16. v. Turkovich, B.F., 1983: On the Metalurgical/Materials, Science Problems in Machining,CIRP ANN. 32, 609-611.
17. Vossloh, M., 1988: Modellgestützte Fehlerfrüherkennung und wissensgestützte Diagnose von Fehlern an Werkzeugmaschinen, Darmstädter Forschungsberichte für Konstruktion und Fertigung, Hanser Verlag.
18. Young, P.C., 1981: Parameter estimation for continuous time models - a survey, Automatica 17, 23-39.

A new integrated concept for condition monitoring and predictive maintenance of machine tools

R. Sandner

Fraunhofer-Institut für Arbeitswirtschaft und Organisation (IAO), Nobelstrasse 12c, D-7000 Stuttgart 80, Germany

Keywords:
Condition monitoring, predictive maintenance, machine modelling, signal processing, knowledge-based systems, scheduling

Abstract:
This paper describes a new methodology for condition monitoring and predictive maintenance of machine tools. It will describe applicable AI-techniques to predict and evaluate the evolution of failures and the degradation of machine tool components. This will lead to find out the causes of occured malfunctions, to suggest corrective and preventive actions and to provide support in the operation mode. The described methodology is expected to be helpful for the improvement of the availability, reliability and maintainability of machine tools as well as for other types of machines or processes.

1. INTRODUCTION

Europe is leading activities in monitoring and diagnostics of machine tools. Real applications to machine tools as well as a great deal of research in tool and process monitoring are taking place and are continuing with a special emphasis on Artificial Intelligence and expert systems. Up to now, most of the work has been done in the field of tool, process and workpiece monitoring [1] and some of these techniques will lead to the machine itself. There has also been a great deal of work in the area of vibration and lubricant monitoring and the application of these to bearings could apply to the machine tool.

Measurement of machine tool parameters for use in diagnostics has taken place in very few laboratories, who have developed model based diagnostics which used the axis drive motor current as the basic measurement to monitor the axis drive, the coolant pump pressure to monitor the coolant system and an acceleration measurement to monitor the main spindle bearings. Only two working demonstrations of machine tool condition monitoring have been found around the world, one in Spain and the other in the United Kingdom [2]. The latter appears currently to be the only place where measurements are used for predictive situations rather than for faults that have already occured.

Because of the black art of the derivation of reliability data for machine tools and the manner in which a condition is decided, condition monitoring is proving a rich area for the application of Artificial Intelligence in the form of neural networks and expert systems. Practically all the work done using Artificial Intelligence is based upon hard fault detection [3]. Hard faults are easily defined but no unified approach including different modelling techniques has been attempted.

There are three different meanings of modelling:
- mathematical modelling, i.e. a mathematical equation, realting some observable variables to a not directly observable variable. The usefulness of such an approach has to be validated by estimating the parameters of the equation from experimental data.
- physical models based on the description of the components and their relations derived from physical laws, describing forces, energy flows, etc. The resulting equations allow the identification of healthy and faulty behavior of the modelled system.
- qualitative models, describing the input and output variables and their changes in qualitative terms with binary figures. Applications have been made for digital circuit simulation [4].

There is a great necessity for a knowledge base containing details of how faults develop and how they exhibit their characteristics. The possibilities for developing such a knowledge base are therefore:
- collecting real data from real machines in industry
- developing a theoretical model by analysis of the failure modes of each element.

Both approaches are part of the integrated methodology described below.

2. AI-BASED CONDITION MONITORING AND PREDICTIVE MAINTENANCE

The methodology for condition monitoring and predictive maintenance describes applicable techniques to predict and evaluate the evolution of failures and the degradation of machine tool components (eg. linear and rotary guides, bearings, ball screw, axis, coolant system, hydraulic, pneumatic, lubrication), in order to determine the causes of occured malfunctions, to suggest corrective and preventive actions. It is part of an ongoing project funded by the European Community under the BRITE/EURAM-programme, and its results will provide improvement in availability, reliability and maintainability of machine tools as well as for other types of machines or processes.

2.1. Requirements of the methodology

This paper desribes an improved way to fulfill the following requirements:
- Information about the actual status of the machine for a given instant of failure has to be acquired.
- The expected status of the machine for a given working point must be estimated.
- Both actual and expected performance, have to be compared.
- If there exist any differences, the causes have to be investigated by identifying the evolution of malfunctions or by diagnosing occurred failures depending on status information.

- Corrective or preventive actions for the discovered malfunctions or failures in an early state have to be suggested.

These requirements result in a basic architectural idea of the monitoring system, containing the three layers "data acquisition", "feature evaluation", and "maintenance planning" (see figure 1).

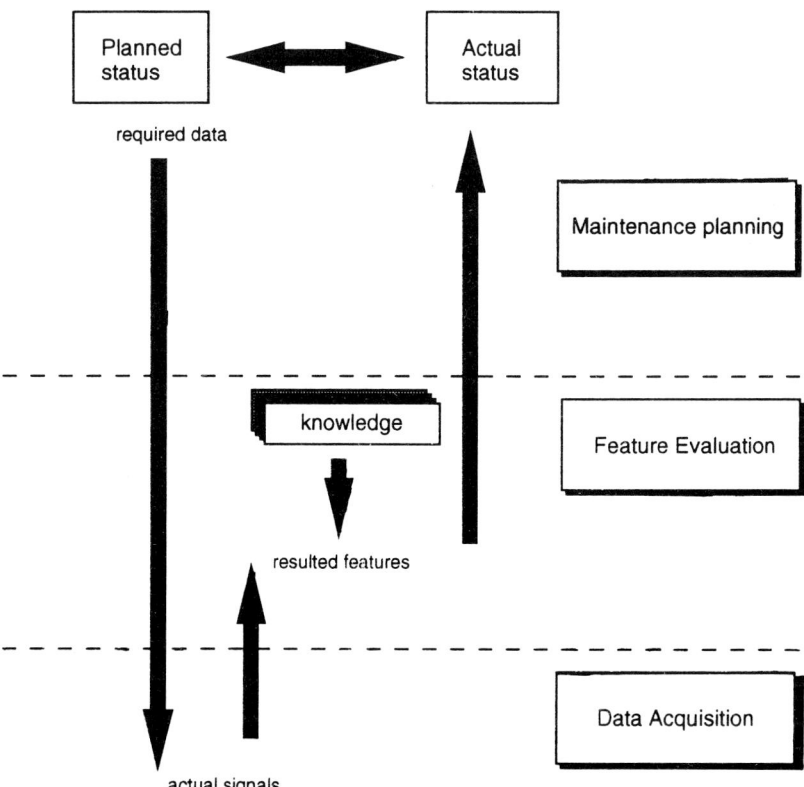

Figure 1. Basic concept of MIRAM

2.2. Modelling of machine tool components

The methodology is based on a detailed description of a three axis vertical machining centre with two auxiliary axis for milling and drilling, its subsystem structure, components, relevant variables, and correlated functional

structure. In order to characterize the selected machine tool several theories and techniques based on:
- physically-quantitative modelling (least-square parameter estimations, instrumental variables parameter estimation, parameter estimation via discrete-time models, limit and trend checking for measurable signals, deterministic approaches like fault sensitive filter models for non-measurable state variables, statistical decision theory)
- qualitative modelling (state diagrams from qualitative calculus, qualitative process theory)

have been assessed and selected for further investigation. A study on the feasibility of different modelling techniques will be performed during the project and will lead to an improved methodology for modelling machine tools in terms of subsystems. Qualitative models are used whenever quantitative ones prove to be inadequate for a PC-based low-cost solution, i.e. by causing unacceptable costs at evaluation time. By the way, the accuracy of machine behavior prediction is also improved by applying qualitative modelling.

Using an approach depicted in figure 2 the refinement of models is driven by the difference between the model and the results of experimental tests carried out at components of the machining centre. Several components have been selected for modelling, depending on the economical interest for predictive maintenance because of high wear-sensitivity, necessity for high availability, unacceptable repair times and costs, or high failure probability. Parameter estimation techniques are used to fit the model to component-related signals and parameters, matching the difference to a predefined criterion. The model will be accepted when the error (i.e. the difference between the parameters predicted by the model and the real ones) is negligible (i.e. it can be considered as noise)

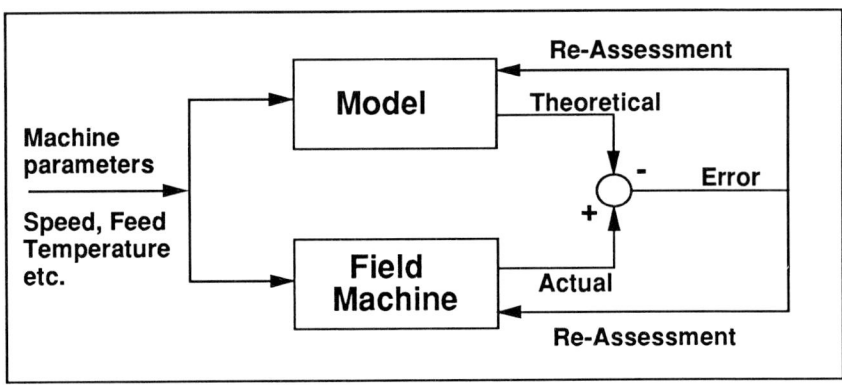

Figure 2. The generation of models

An existing difference produces either a change in the model through continuous parameter adjustment or requires a modification of component related measuring points by adding new sensors (eg. ultrasonic sensors for measuring impact sound to improve wear prediction).

The component models serve as an important starting point for further elaboration of condition monitoring and predictive maintenance, which include the development of a fault dictionary and a knowledge based evaluation of a given machine subsystem status. Thus, intensive simulation, based on the models is considered as the necessary second step to obtain relevant fault information in terms of related input and output vectors.

2.3. Model-based failure simulation

Existing digital simulation techniques have several advantages like
- applicability to calculations of reliability and availability
- a high mapping exactness with a quite simple theory lying behind
- coverage of a large variance of different monitorable conditions.

The behaviour of machine tool components are simulated through so-called "macro instructions", whereas the component characteristics may be represented as interactive software modules which communicate with other components using messages. Failures of machine tool components could then be produced through a random time generator. Nevertheless, this approach has certain utilization limits. If one replaces the random number generator by a function taking into account statistical failure occurences as a result of a reliability analysis, carried out on the selected machine tool, a serious advantage is reached.

In the methodology described here failure simulation is used to relate output variables to input parameters. Every component model is considered as a black box, to which a variety of values of input parameters and signals is presented as input vectors, and a single output vector is produced as a result after each simulation run. These output vectors are compared to identify significant differences, which are considered as an indication of a faulty component state. Statistical signal detection techniques like chi-square estimation or ANOVA are performed in order to ensure that the detected differences actually indicate an observable faulty state, and are not simply Gaussian distributed noise. The output vector, indicating a faulty state, is mapped to all input vectors to identify the related subset of them.

These tupels of input/output vectors are used to construct or update the failure characteristics in the fault dictionary.

2.4. Knowledge based failure prediction using a fault dictionary

The fault dictionary, based as well on results of experimental tests at the machine and on simulation results, contains a full description of failures with related signals (coolant system temperature and pressure, lubrication flowrate, motor current, speed, rpm, forward feed, command voltage, spindle vibration, bearings acceleration, oil consumption, etc.), related components (axis, spindle, coolant, bearings, ball screws, guides, etc.), input parameters (risetime, rms, cycle time, etc.), failure probability and repair advices. It is the main input for the knowledge base, used for health monitoring and fault diagnosis, and will be periodically updated either by evaluation results of an actual subsystem status or by simulation (see figure 3). Thus, the fault dictionary provides all necessary input to construct a diagnostic knowledge base.

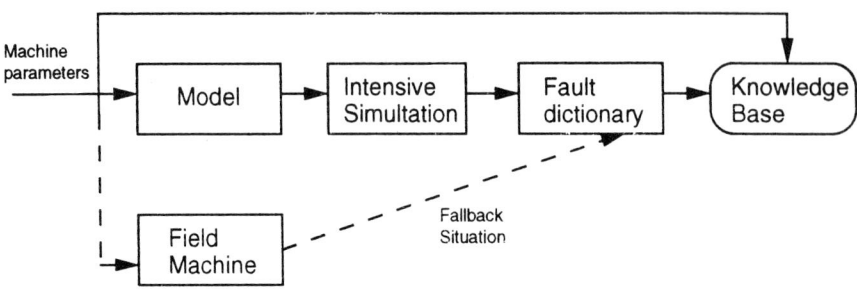

Figure 3. Generation of knowledge bases

The application of knowledge-based techniques for diagnosis of soft and hard faults is an useful approach to evaluate the status of machine tool subsystems. There exist a lot of projects dealing with knowledge-based systems for machine tool diagnosis. Current systems focus mainly on hard faults for example in electrical or hydraulic machine tool components. In the project described in this paper special emphasis is put on the prediction of soft faults as soon as possible and on the early recognition of malfunctions. This brings up several requirements on the diagnostic capabilities:

Diagnostic techniques have to trace back on failure characteristics and the component description represented in the fault dictionary in order to predict failures based both on features extracted from actual process signals and the estimation of unmeasurable parameters and state variables (e.g. prediction of wear and tear). Previous diagnosis of failures was restricted to checking directly measurable variables for upward or downward trangression of fixed limits or trends. Various faults of machine tool components could then be detected, but only after the measurable values have changed. Within this project an improved methodology for knowledge-based diagnosis is developed in order to enable the use of advanced techniques which can detect particularly soft faults earlier and which can locate them better.

In order to evaluate features extracted locally at the data acquisition elements, the methodology for prediction and diagnosis is slightly different to well-known rule-based diagnostic strategies. The features observed are initially processed by pattern recognition facilities, using pattern classification by likelihood functions and a statistical approach for trainable pattern classifiers.

Once a feature has been recognized as a significant indicator for a faulty state, it is established as a key feature for further processing. This process uses information about involved parameters and signals and their evolution, which are related to a malfunction, and match it against the observed key feature in order to predict or detect a failure. The detection of hard faults is based on knowledge about components and related malfunctions, assuming that a hard fault is definitely identified, if the causal chain of related signals and parameters can be matched to a malfunction unambiguously. The prediction of soft faults depends on the availability of trend characteristics for both healthy and faulty component behavior. Once a feature has been determined as a key feature, it has to be decided, using appropriate statistical significance tests, whether it represents the normal or one of the faulty trends. Depending on the time characteristics of the identified trend curve, prediction of the future evolution of the component state is possible. An overview of this process is given by figure 4.

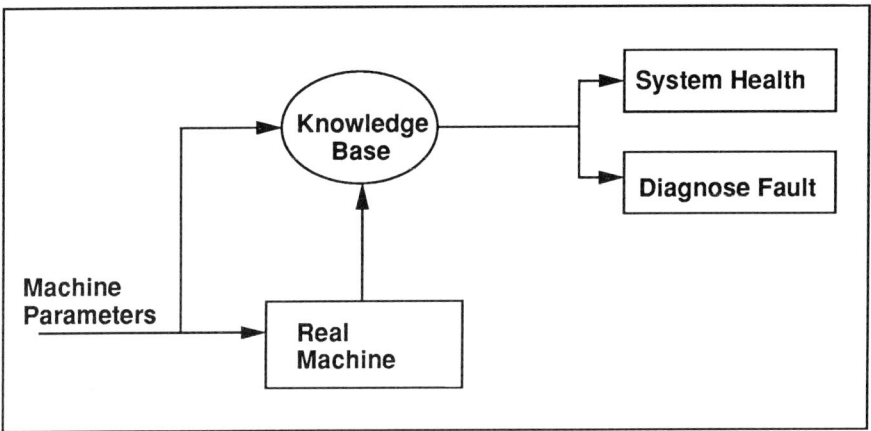

Figure 4. The process of condition monitoring and fault detection

As the fault dictionary provides the knowledge base with all available information about failure characteristics (risk failures related to components, failure probability, trend curves, faulty state vectors, etc.), the implementation of a real-time access to the fault dictionary will be a preliminary step in implementing a prototypical knowledge base.

Another requirement of failure prediction is the capability of dynamic reasoning functionality. Whenever a new unknown feature arises, which cannot be recognized by the pattern classifier, an adaptation process to refine the knowledge base by updating the fault dictionary becomes necessary. We assume that the correct recognition of unknown features is a stochastic process with denumerable states with probabilistic transitions, thus learning based on Markov models is the major paradigm used for dynamic reasoning.

There are several requirements on the adaptation process which are specified as following:
- Criteria for similarity between unknown and recognizable feature must drive the assessment of new features.
- If a feature is unmatchable, it is necessary to determine why, i.e. determine which signals and parameters are significantly different.
- The value of similarity has to be estimated by comparison to expected templates in the fault dictionary.
- The causal relationships between value changes have to be infered by looking at the component structure to identify likely affected components.
- Criteria for deciding if a new feature will become an updating entry to the fault dictionary have to be elaborated (e.g. using maximum-likelihood estimation to assess the cost of an entry vs. the risk of a non-entry).

The knowledge obtained in this way is complemented by the heuristic knowledge provided by the machine designers and operators.

2.5. Maintenance planning

Once a failure has been detected or predicted, corrective or preventive actions taking into account results from diagnosis as well as from simulation are suggested to the machine tool operator or the maintenance personnel. Thus, strategies for maintenance planning are integrated into the overall methodology. They are mainly based on the following input:
- the detected occured failures or the predicted malfunctions
- the availability of maintenance operations
- relevant economic or logistic constraints of the manufacturing process (e.g. repair costs, available resources and material, stopping costs, criticality of production time)

In the machine monitoring environment these corrective or preventive actions are based on a multi-step planning process that is generated and verified by simulation. Plans are repair strategies, designed to bring the machine to the planned status of operation as close as possible. If this is not possible and the failure is time-critically related to the actual operation, emergency actions (shutdowns, stops) are performed to avoid additional damages.

Two basic approaches for planning maintenance actions are integrated in the methodology. In the classical one, plans are constructed on demand out of a set of plan primitives. The second approach consists basically of making use of plans generated in the past, and is known as case-based planning. Here the planner exploits its own experience, recalling past plans for similar situations and modifying them to suit exactly the actual needs.

In the context of case-based planning, many of the disadvantages of classical planners are avoided:
- Rather than planning for individual goals and then merging the results, a case-based planner searches its memory for plans that satisfy as many as possible of its goals in parallel.
- Rather than discarding the plans it builds, a case-based planner can save them in memory for later use in similar circumstances.

3. CONCLUSIONS

In order to implement the methodology described so far in the real environment, a data acquisition system, gathering signals and parameters, which describe the status and the behaviour of machine tool subsystems, is currently developed at one of the project partners in Wales as a first step for the machine monitoring prototype. The overall functionality of the data acquisition system depends on the modelling of the machine as described above, the availability of existing sensors and new ones, if required by parameter choice. Special emphasis is put on the sensitivity of the signal parameters to different faults.

Features from acquired signals and parameters are extracted by applying the following analysis techniques among others: Fourier analysis of periodic axis or spindle vibrations, analysis of cycle time, dynamometry, analysis of process parameters like pressure, flow, temperature, etc., over time in order to identify tendencies, steady state and transient profiles.

The overall methodology for condition monitoring and predictive maintenance of machine tools will be implemented on a fully functional prototype in order to determine its feasibility in real industrial environments. This implementation will take place step by step, which allows continuous refinement depending on results arising during later project phases.

The project is expected to be finished in middle 1994 and to deliver a new approach for the diagnosis of hard faults and the expansion of the prediction of soft faults to wear parts inside the machine tool itself. The methodology developed for these goals is also applicable to other kinds of machines as well as to extended processes (e.g. FMS, assembly line).

4. REFERENCES

1. J. Tlusty, G.C. Andrews: A Critical Review of Sensors for Unmanned Machines. In: Annals of the CIRP Vol. 32/2/1983
2. K.F. Martin, P. Thorpe: Coolant System Health Monitoring and Fault Diagnosis via Health Parameters and Fault Dictionary. International Journal of Advanced Manufacturing Technology, 5, 1990, pp. 66-85
3. R.D. Puetz, R. Eichhorn: Expert Systems for Fault Diagnosis on CNC Machines. International Journal of Man-Machine Interaction, 26, 1987, pp 87 - 96
4. T. Tanaka: Structural Analysis of Electronic Circuits in a Deductive System. In: L. Bolc, M.J. Coombs (Eds.) Expert System Applications. Berlin, New York Heidelberg, 1988

NEW PHYSICAL PRINCIPLES IN MANUFACTURING TECHNOLOGY

CHAIRMAN: D. KOCHAN
TECHNICAL UNIVERSITY DRESDEN, GERMANY

Stereolithography - Fields of Application and Factors Influencing the Accuracy

Prof. Dr.-Ing. B. E. Hirsch; Dipl.-Ing. H. Müller

BIBA, Bremer Institut für Betriebstechnik und angewandte Arbeitswissenschaft an der Universität Bremen, Klagenfurter Strasse/Betriebshof, 2800 Bremen 33, Germany

Abstract

The article gives an introduction to the new "Rapid Prototyping" techniques. As an example for these manufacturing methods the basic principle of Stereolithography (SL) is described. The fields of application, the present limitations and the model accuracy of this method are mentioned. The factors influencing the accuracy of SL model are outlined and the main factors described in more detail.

1. AN INTRODUCTION TO NEW RAPID PROTOTYPING TECHNIQUES[1]

A range of new manufacturing techniques, which have generated great interest for some time now, promise a significant improvement in the present situation in production of one-of-the-kind products, prototypes and their tools. The ideas described are so innovative that the possible consequences for the manufacturing industry can be compared with the development and wide-spread use of the NC technique. These new techniques are variously called "Rapid Prototyping (RP)", "Freeform Manufacturing", "Layer Manufacturing Techniques" or "Desktop Manufacturing". These names have still not been clearly defined.

These new techniques hope to give a tool to designers and engineers enabling them to produce models and prototypes with a minimum of human interaction directly from CAD data within a few hours, in an office environment where at present only printers and plotters are being used. RP could be of great importance for Industry. It will speed up the design process, thereby contributing to reduced costs while allowing for the production of more complex and sophisticated products.

At least seventeen RP processes are currently known. Their level of development ranges from the genetic, to first prototype, to models at the stage of market introduction, right up to units with years of successful industry committment. This explosive rise in the range of concuring ideas gives a very clear indication of the world-wide future potential of this innovation.

---

[1] Adapted from /11/

1.1. The Basic Principle of Stereolithography

Stereolithography is one RP-technique and at present it is the only one widely used in industry. About 200 systems are in operation worldwide, 50 in Europe (Autumn 1990). The original idea for the process dates back to 1982. It was first publicly demonstrated at the AUTOFACT exhibition in October 1987. It was developed and is produced by 3D Systems Inc. in California, USA. Stereolithography complements CAD, laser and polymer technologies to form a new technique for the production of 3-dimenisonal plastic pieces.

The principles underlying the construction of a 3D object are shown in figure 1-1. At the start of the process, the platform within the chamber is positioned one slice-width, below the surface of the liquid photopolymer. An X-Y scanner guides the laser beam over the liquid material and, using the vector-pattern generated by the "Slice calculator", draws out the first slice. The liquid plastic is polymerised on contact with the U-V laser beam. When the first slice is finished, the carrying platform is lowered by one layer. The next slice is then processed and simultaneously connected to the previous one. In this way the workpiece is gradually built up in slices, working from the bottom up to the top.

When the last slice is finished the piece is taken from the chamber. It is cleaned, and the supports are removed. After this the model is placed in a U-V lightbox to be completely hardened. This step is necessary as the laser-beam alone does not fully harden the resin in the chamber, but rather leaves a quasi-laminate structure, within which still liquid material can flow. After being fully polymerised in this post-process, the workpiece can be further processed in more traditional ways. Lacquering, coloring or coating are also possible.

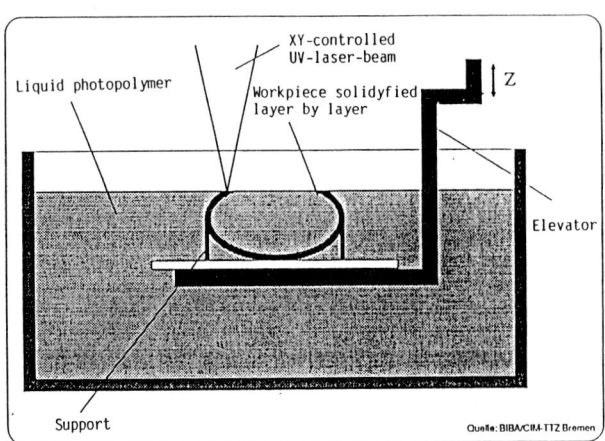

Figure 1-1: The principles underlying the construction of a 3D object

2. FIELDS OF APPLICATION AND PRESENT LIMITATIONS OF STEREOLITHOGRAPHY[2]

Just a few hours after the completion of the construction process a finished model is availabe. This is the main advantage of the technique and a definite improvement over the current situation, where models and prototypes are painstakingly built. The more geometrically complex the object, the greater the potential savings both in time and money. The use of stereolithgraphy is especially suitable when there is a need to quickly convert ideas into "ready-to-hand" pieces. When one weighs up the two main advantages, savings in time and money, the saving of time is the more important of the two.

Stereolithography has aroused much interest in Industry, mainly because of its usefulness in three particular areas.

2.1. Models for "Show and Tell" or "Touch and Feel" Purposes

Refinements in computer simulations and displays still do not obviate the need for a life-like illustrative model. If a product is to be prepared and produced such models are absolutely necessary. It is no surprise, therefore, that stereolithography is of great interest to designers and development engineers. A designer may wish to look at a model of a perfume bottle or a telephone handset. The development engineer can check the grip of a new tool. Presentations to clients are made clearer and more attractive when the firm can present a model and not just a series of picture in order to explain the potential of a new product. Real models also improve the working relationship with suppliers, and enhance co-operation among groups simultaneously working on a project. They help toolmakers, for example by accurately comprehend the complex shape of a rubber sealing of a car. More exotic examples are the modelling of molecules, or the reproduction of a small piece of an oil containing mineral, or even the exact modelling of a part of the human body prior to a complex operation.

In the production of realistic 3D models for presentation and technical co-operation, stereolithography has fulfilled all expectations. This is currently the most important use of the technique.

2.2. Functional Model

Before any product can go into full production, his functional capacities and reliability must be proved. The detail of testing depends on the complexity of the final product. For example:

- Because of the high complexity of the cooling pipes it is difficult to predict whether, for instance, an electromotor will receive sufficient cooling ventilation. Tests using models are then necessary.
- The rubber tubing of an automobile is tested for ease-of-assembly and practice correctness within a simulated environment.

---

[2] Adapted from /11/

Presently, pre-production examples are made for these test. The demands placed on there are less than on the final product, as time and money considerations are important. Usually 5 to 30 pieces are needed, whose dimensions correspond closely to those of the final product. They should be from the some material, or at least a material with closely comparable properties.

The application of stereolithography in this field could open the way to great savings. However, two factors limit the use of currently available SL-machines in this area. First, the materials used by SL-machines are still too different from those of final plastic products. Secondly is the dimensional accuracy not good enough.

2.3. Stereolithography Models As A Basis For An Other Subsequent Process

Stereolithography is best at producing complex geometric forms. The CAD systems are able to provide information to the SL-machine on the "positive" or "negative" shape of the workpiece. The SL-model can be used on the production of a copy from some other material using special manufacturing techniques.

Potential applications are:
1. The production of moulds; this is still in practice possible only with great difficulties. The material is neither solid nor thermally stable enough. The dimensional accuracy is insufficient, and unforeseen warpage tend to occur.
2. Production of patterns for sand and investment casting.
3. Production of master models for metal spraying, epoxy tooling, or vacuum casting.
4. Production of master models for die casting or EDM electrodes.

Work is ongoing on all potential applications, but is still at the research and development stage. The two main problems to be overcome are the properties of the SL-material and the inadequate accuracy of the technique for this kind of application.

3. THE ACCURACY OF THE STEREOLITHOGRAPHY PROCESS TODAY

Depending of the size of the workpiece, a tolerance of between $\pm$ 0.25 mm and $\pm$ 0.5 mm can be expected. It is also possible to achieve closer tolerances, but this however requires careful tweaking of the parameters, repeated runs of the process, and a little bit of luck. In this painstaking way a tolerance of $\pm$ 0.15 mm is possible.

This very general statement has been confirmed through measurements taken on 22 identical workpieces. The parts, built on 19 different SLA-250 machines, were compiled at the 2nd European Stereolithography Users meeting (October 1990). The Slice-parameters for the support were given, those for the parts were chosen by the users. The part is shown in figure 3-1.

Table 3-1 shows the minimum, maximum and average values, and the standard deviations for the measurements. The standard deviations are of the greatest interest in the light of this discussion. The measurements of x1 and y1, both with a numerical value of 30 mm, have standard deviations of S_x = 0.141 mm and S_y = 0.132 mm.

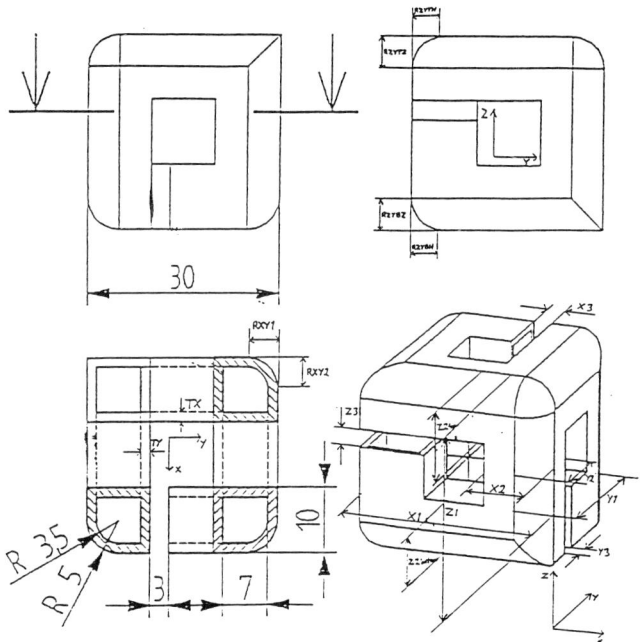

Figure 3-1: The Demo-part with actual measurements

	Mod. dim.	Min	Max	Average	Std. Dev.
X1	30,00	29,63	30,23	29,87	0,1413
X2	10,00	9,86	10,24	10,07	0,0861
X3	3,00	2,58	2,91	2,74	0,0855
Y1	30,00	29,71	30,22	29,94	0,1321
Y2	10,00	9,79	10,22	10,07	0,0962
Y3	3,00	2,42	2,82	2,61	0,0922
Z1	30,00	29,55	30,62	30,08	0,2398
Z2 - top	10,00	10,11	10,57	10,30	0,1503
Z2 - bottom	10,00	9,86	10,82	10,16	0,2180
Z3	3,00	2,24	2,85	2,58	0,2017
TX	1,50	1,59	1,95	1,74	0,0864
TY	1,50	1,60	1,85	1,73	0,0755
TZ	1,50	1,42	2,08	1,67	0,1786
RZYBH	5,00	3,46	13,76	6,01	2,5736
RZYBZ	5,00	4,50	5,88	5,09	0,3702
RZYTH	5,00	3,28	4,67	3,82	0,3342
RZYTZ	5,00	3,99	5,78	4,86	0,4364
RXY12	5,00	4,44	6,04	4,99	0,3601

Table 3-1: Statistics concerning all measurements of each and all Demo-parts

According to the statistical theory, this means that about 95% of measurements will be in a tolerance band of $\pm 2S_x = \pm 0.282$ mm and $\pm 2S_y = \pm 0.246$ mm or 99,7% in a range of $\pm 3S_x = \pm 0.423$ mm and $\pm 3S_y = \pm 0.396$ mm.

High deviation and the incertainty to predict the result limit the application significantly. This has also been recognized by the manufacturer.

In the past two years the Process Department at 3D Systems has studied a number of the most important sources of error in Stereolithography. As a result of this research two improvements with respect to accuracy have been developed:
- A new building method known as WEAVE
- The use of flourescent lamps in the post-curing process

The intent to WEAVE was to reduce the fraction of liquid resin remaining within the laser cured part, while simultaneously attempting to minimize curl distortion. To accomplish this, a drawing method was developed involving orthogonal vectors with two passes per layer. Specifically, a set of parallel Y hatch vectors followed by an orthogonal set of parallel X hatch vectors. However, in a departure from previous build methods, the first set of cured hatch lines are intentionally separated by about 1 mil (0.025 mm) from each other. The hatch lines are also intentionally given a cure depth about 1 mil less than the layer thickness. Hence, the first set of hatch vectors neither touch each other nor the cured layer below /12/. WEAVE should significantly reduce postcure distortion.

The use of flourescent lamps instead of mercury-vapour lamps shall also contribute to reduce postcure distortion. The new techniques were tested with special testpieces. Their results are promising. Measurable effects at models built by users are not yet known.

4. FACTORS INFLUENCING ACCURACY

All information given in the following section refers only to the machines SLA 250 and the resin XB 5081-1 (brittle raw material). The linear dimensional tolerances will be discussed.

Other groups of tolerances such as ovality, concentricity angles, radii or suface finish are not discussed.

Many elements within the process affect on the final tolerance of the model, and these various factors are important at different stages of the process, where they have various effects.

Table 4-1 shows the more important factors which influence the accuracy of the final product. They can be classed into the following four groups:
- Data Processing
- Scanning System
- SLA 250 mechanical systems
- Building accuracy and material behaviour

In the section "tolerance field" the tolerances are discussed. The discussion is based on
- Specification of the manufacturer

ELEMENT GROUP	ELEMENT	TOLERANCE FIELD	Affected directions	COMMENTS
Data processing	Facet representation:	±D1	XYZ	
	Stair Stepping Effect:	+D2	XYZ	
	Beam Width Compensation:	±0.01mm	XY	Data processing.
	Units of Measurements and Shrink Factor Calculation:	±0.01mm	XY	CAD-units → Slice-units → Mirror-units. The new value must fit in the Mirror unit grid.
Scanning system accuracy	Spot Location Accuracy:		XY	Dependent on where a part is positioned in the working area.
	Spot Location Repeatability:	±0.06mm	XY	
	Part Position in the Building Process	±0.01mm	XY	
SLA-250 other mechanical systems	Elevator Positional Accuracy:	±0.06mm[1] ±0.013mm[2]	Z	Is also affecting the over cure depth.
	Resin Surface Positional Accuracy:	±0.03mm	Z	No tests have been done. ±0.03 is the claim of 3D systems. Is also affect. the over cure depth.
Building accuracy and material behaviour	Line Width Alterations:	±0.06mm	XY	Due to both the elipticality of the laser beam spot and different cure depths within a part.
	End Points of Hatch Vectors on Border Vectors:	±0.02mm	XY	Hatch vector- and hatch spacing dependent.
	Over Cure on Down Directed Surfaces:	OC +0.20 ±0.09mm	Z	OC is the over cure that can be read in the Prepare Menu. The tolerance range is valid for parts with three hatch directions.
	Inhomogeneous Shrinkage in the Building Process:	±0.14%	XY	The 0.14% (of the dimension) value is valid for parts having X- and 60°/120°-hatch with 1.00mm h.-spacing. The value is also affected by the elipticality of the laser beam spot.
	Inhomogeneous Shrinkage in the Post Curing Process:	±0.05mm	XY	
	Influence of Cleaning Process Step			
	Curling, Warpage and Edge Curling:		XYZ	Dependent on part geometries, building- and post curing parameters. Deviation can be highly considerable.
	Swelling:	±0.05mm	XY	Dep. on part geometries and building time.

1) Vertical accuracy; Spectra Physics Specification 3D 109
2) Position repeatability; specification 1990 3D Systems, Inc.

Table 4-1: The significant elements affecting the accuracy in SLA building process

- Completed tests to isolate the different influencing factors
- Experiences gained from use of the machine at BIBA

This section is an attempt to quantify the current level of knowledge and experience. The table should give the user an indication of the influence of the process steps and machin parts on workpiece accuracy, and explain the order of magnitude of the associated tolerance fields. These results are not yet scientifically proven in a strict sense, in particular those relating to "Building accuracy and material behaviour". It must be said that the influences on accuracy which are discussed here in most cases depend on several factors of which the interaction has to be more investigated. Now the influencing groups will be discussed in more detail.

4.1. Data Processing

This group includes the influencing factors associated with the following process steps:
- CAD/Stereolithography
- Slice
- Stereolithography data processing.

When the CAD internal format of a model is converted to the planar representation, the STL-format, the first imprecision worth noting is introduced.

Due to the fact that curved surfaces in the STL-format are represented by a limited number of plane triangles, it is obvious that the stereolithography data format can only offer an approximation of the shape of the object. The deviation introduced is dependent on the particular radius of the curvature, the size of the triangles, and whether the surface is convex of concave.

Slicing of the model creates a "stair stepping" effect. The steps are always outside of the faceted object and thus contribute with a positively signed deviation.

The slice software includes a function to adjust the laser inward in outer boundaries, and outward on inner boundaries, so that the diameter of the laser beam is compensated. This function should always be used. If not it will result in a deviation of about $\pm\ 0.25$ mm. In the chain of data processing steps involved in stereolithography a number of different units are used. When a number is translated from one unit of measurement to another one, the number must be rounded to fit within the new scale. Within the stereolithography process this rounding occurs twice, from CAD- to slice- and slice- to mirror units.

4.2. Scanning System Accuracy

One test method to isolate errors resulting from scanning system accuracy is to etch with the laser beam directly onto a photographic film.

However, there are several obstacles to overcome before a satisfactory test can be performed:
- The photographic film must have an appropriate photosensitivity. The laser beam is much more intensive than light used for normal photography. It is also desirable to be able to use a laser beam drawing velocity comparable to the velocities used for practical building purposes.

- The photographic film has to represent the laser beam's scanning lines as accurately as possible. In other words; the film must have a high resolution and the reflection of the laser beam (fluoresence) must be low.
- There must be no shrinkage nor warping of the photographic film, neither during the exposure nor during developing.
- The photographic film must also during the exposure be positioned perfectly horizontally and, simultaniously, at the exact height of the resin surface.

Tests with photographic paper (ILFORD Photographic paper) showed that this material was not suitable for this purpose. Hologram plates were then used. This material proved to be adequate to the task. The type 8 E 56 HD from AGFA-GEVAERT was chosen. It has a very high resolution and a low photosensitivity which makes it possible to apply practical drawing velocities for the laser beam. The hologram plates are available in the two sizes 5"·4" and 10"·8".

4.2.1. The Laser Beam Spot

Size and shape of the laser beam influences directly the accuracy of the models. The diameter of the spot is a contributory factgor both for the cure depth and the cured line width. The influence of the shape of the laser spot can be directly compared with the influence of the geometry of a milling tool. In order to get an answer to this question, a test with hologram plates was performed.

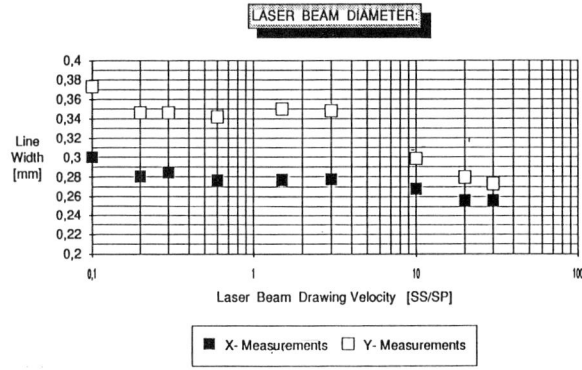

Figure 4-1:   Laser beam diameter test

Nine line pairs, each pair with lines in X- and Y-direction respectively, were drawn with different drawing velocities on a hologram plate. The plate was positioned in the middle of the elevator platform at resin level hight. The width of the lines were measured at two places on each line, and the average value was plotted into the chart shown in figure 4-1.

As can be seen from the chart, there is a significant difference between the line widths in X- and Y-direction. In other words; the laser beam spot in the BIBA's SLA-

250 was not circular but eliptic at this time, with a bigger diameter in Y-direction than in X-direction.

It is difficult to exactly estimate the diameters of the spot from the line widths on the hologram plate. However, the fact that the X- and Y-measurements vary very little with a relatively wide range of laser beam drawing velocities indicates that the diameters can be read to a high degree of accuracy from the chart. It can thus be concluded that the diameter in X-direction is about 0.28 mm, while the diameter in Y-direction is close to 0.35 mm. This gives an X- to Y-diameter ratio of 0.80.

Figure 4-2 shows how the intensity of the laser beam spot varies respectively in the X- and Y-directions. The results of the beam analysis using the beam profilers have been plotted here. At the Y-direction boundaries the light intensity has a lower level compared to the X-direction boundaries. This explains the decreasing Y-line width when drawing with very high drawing velocities. The applied energy of the laser beam spot onto the holographic plate is then insufficient to blacken the film.

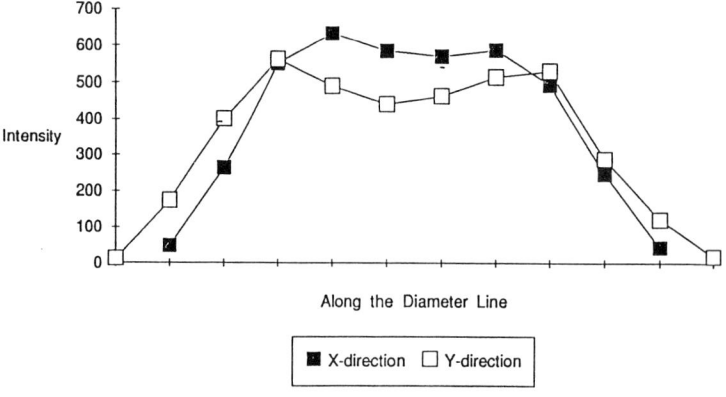

Figure 4-2: Intensity distribution along the X- and Y-diameter line

Figure 4-2 also shows distinctly that the laser beam spot in the BIBA's SLA-250 has a greater diameter in Y-direction than in X-direction.

It is, however, interesting to take note of the results from a beam focus examination on the BIBA's SLA-250, carried out by the 3D systems' service person the 14th of March 1991. This examination showed an X- to Y-diameter ratio of 0.87 and a diameter of the laser beam spot of 0.19 mm. The laser power was at the time only at about 6mW, and the beam profiler's accuracy and ability to register such low intensities is not known.

Finally it is pointed out that size and shape of the laser beam are machine dependant parameters which should be checked by each user individually.

4.2.2. Spot Location Control

In order to check the spot location control the grid pattern shown in figure 4-3 was drawn on three different 10" x 8" hologram plates.

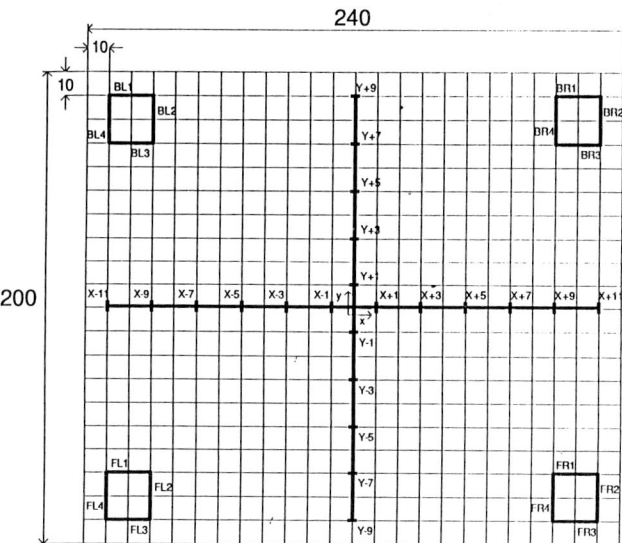

Figure 4-3: Grid pattern with measurements

The coordinates of the points shown in the figure were measured on the plates. The following general conclusions can be made:
1. If one measures the distance between two measuring points, the standard deviation in the measured distance is circa 0.02 mm. This means that one can say that 95,4% of all results will be scattered in a region of width of ± 0.04 mm (2 x standard deviation, or 99,7% within a range of ± 0.06 mm (3x standard deviation). Details of positional accuracy are not provided by the manufacturer.
2. The distance BR1-BR2, BR3-BR4, (X+9)-(X+11), FR1-FR2 and FR3-FR4 were on all the three hologram plates significantly bigger than other corrosponding measurements (the actual range is from about 0.02 mm to 0.12 mm). This means that the mirror units in the X-direction, in the very right region of the working area, are bigger than the mirror units elsewhere. In other words, the mirror unit grid is not uniform over the whole working area.
Apart from this region to the right of the working area, there are no repeatedly significant differences in the X- and Y-directions.
3. In order to check the spot location repeatability, the measuring results of two hologram plates were subtracted.
The result is charted in figure 4-4. The deviation ranges from -0.027 mm to + 0.044 mm. This range is within the spot location repeatability of ± 0.06 mm which is the value given by the manufacturer.

Finally it is mentioned again that the manufacturer's specification don't tell the accuracy of positioning. The specified spot location repeatability of ±0.06 mm is not good when compared to the values of NC machine tools. A value of ±2 increments of the measuring system is standard here, which means a repetitive accuracy of at least ±0.02 mm.

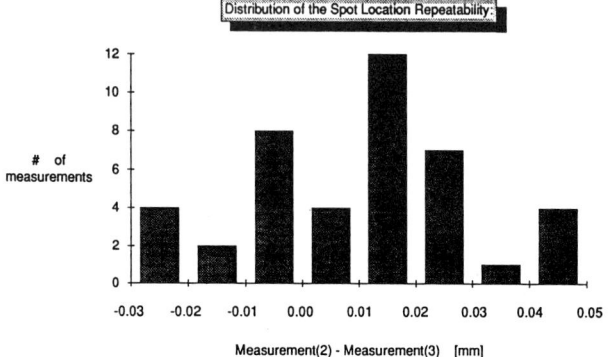

Figure 4-4: Spot location repeatability

4.3. SLA 250's Other Mechanical Systems

This chapter deals with the elevator positional accuracy and the resin surface positional accuracy.

The first was tested by the Danish Technological Institute in Arhus to which we will be referring /8/. They found a repetitive accuracy of ± 0.01 mm and an accuracy of positioning of ± 0.04 mm (for a distance of up to 0.3" which equals 7.61 mm). Two different values are contained in two different specifications of the manufacturer. One older spec. (Sp.-Ph. 3D 109) says "vertical accuracy ± 0.06mm", the second spec. (1990 3D Systems, Inc.)" position repeatability ± 0.013mm".

The weight of the Stereolithography parts on the platform will not affect the accuracy in the Z-direction to any remarkable degree (the pressure to the platform is very low due to the buoyancy).

The SLA is using a laser measuring system to measure the resin level. The surface positional accuracy, as well as the surface positional repeatability, is said to be ± 0.03 mm.

4.4. Building Accuracy and Material Behaviour

This group contains the various influences on the final product accuracy arising from material quality, work-piece specific building parameters, or from process-specific effects (curing and post-curing). Combined, these parameters exert the greatest influence on the final product quality. Their individual influences are shown in tabe 4-1.

Distortion of the model as a consequence of material shrinkage is the most dominating effect. As it is not possible to fully discuss all the various factors here, we concentrate on a general statement concerning this material property.

Shrinkage is occuring in two stages, in the building process on the machin and in the post curing step.

Shrinkage in the building step is influenced by the scanning method. The direction of the different hatch vectors and the sequence in which the hatching is drawn on the surface of the resin are of importance.

At BIBA many testpieces were analysed in this direction. All parts had the same hatch parameters, x and 60°/120° hatch with 1 mm hatch spacing. The shrinkage in y was a little larger than in x direction. This result can be explained by the sequence in which the vectors are drawn (at first 120°, then 60° and finally the x direction) and the larger number of intersection points on the 60°/120° vectors when the x vector is produced. These intersection points will thus undergo a further shrinkage which will affect the measurements in y-direction more than in x-direction. Further the over cure is greater when the laser is moving in y-direction due to the shape of the laser beam (at BIBA's SLA-250 at the time when the tests were made). This effect gives a greater addition to the shrinkage in the y-direction.

An other important parameter is the amount of liquid material inserted in the model. It depends on the density of the hatching. It was observed that the post cure distortion is proportional to the fraction of the liquid remaining within a laser cured part /12/.

Experimental observation also showed that curl distortion (the bending of multiple unsupported cantilever layers) is related to the amount of polymer shrinkage occuring after contact with the previous layer /12/.

The experience when building models shows that shrinkage is closely related to the model geometry. Tests showed that thin walls shrink more expressed in percentage than thick walls.

In the foregoing single influencing factors were mentioned. Unfortunatelly it is at the moment not yet possible to predict their impact on the final part quantitatively. Here still much research work has to be done.

5. SUMMARY

The accuracy of the final Stereolithography model is determined by a chain of influencing factors.

It is to a high degree dependent on the SLA operator. He has to apply accurately and carfeully the appropriate Slice- and Building parameters. Due to the fact that in particular the Prepare Menu interface is absolutely user unfriendly, the Range (.R) file always has to be checked before building (although this is eliminated by Software Release 3.81). Further, the focus of the laser beam spot and other important hardware modules must regulary be checked.

It is also emphasized that several influencing factors will vary from machine to machine in such a way that they are worth taking into account when choosing the slice and building parameters. One such parameter is the shape of the laser beam spot, another can be the accuracy of the scanning system.

The way in which the elements in the tolerance chain affect the dimensional precision varies. Some elements vary within absolute ranges, but there are also elements that vary within ranges, which are dependent on part dimensions and geometry. Again, other elements are interdependent with process time considerations in the Stereolithgraphy process.

Also the post treatment and precise finishing can significantly improve the final tolerances of Stereolithography parts. The information which the user gets from the manufacturer, concerning this process step, is too poor.

Finally the results given in table 4.1 are compared to those of the Demo Parts described in chapter 3. A tolerance band of ± 0.246 mm to ± 0.282 mm for 95 % of the measurements, respectivly ± 0.396 mm to ± 0.423 mm for 99,7 % of the measurements was found out for nominal values of 30 mm.

When adding up the tolerance fields in the X- and Y-direction given in table 4.1, a final range of $[\pm (0{,}27 \text{ mm} + 0{,}14\% + D1_{XY} + D2xy]$ appears. For a nominal value of 30 mm we get a tolerance band of ± 0.31 mm. However, with respect to the theory about standard deviation of a sum, the real tolerance range will be smaller. How much smaller is not yet known.

For the moment we can say that both results lie in a comparable range.

6. ACKNOWLEDGEMENTS

This paper is possible due to the funding which BIBA recieves from in the European Research and Development Project BRITE/EURAM (Project INSTANTCAM, Nr. BE-3527-89). The authors would like to thank the Directorate "Technological Research" of the Commission of the European Communities as well as the ten partners from five countries for their generous assistance. They would also like to thank Leif Störmer for his excellent Diploma work.

7. REFERENCES

/1/ "National Conference on Rapid Prototyping"
June 4-5, 1990, Stouffer Center Plaza Hotel, Dayton, Ohio.

/2/ "DESKTOP MANUFACTURING - The next automation revolution"
Published by Technical Insights, Inc.
Englewood/Fort Lee, NJ, USA

/3/ "INSTANTCAM"
BRITE/EURAM Project No. BE-3527-89
Workarea 1, Workpackage 1+2, Paper 1.
Title: Study and development of different techniques, machine concepts and materials.

/4/ "Computer Aided Manufacturing of Three Dimensional Objects"
J. C. Andre, S. Corbel, F. Nonnenmacher, P. Schaeffer.
GRAPP-URA 328 CNRS AND GDR 0920 CNRS, ENSIC - INPL, BP 451, 54001 NANCY CEDEX, FRANCE.

/5/ "Investment casting - actual technology and tolerances"
Mechanical Engineering Department of Instituto Superior Técnico, Portugal
INSTANTCAM, Work Area 5

/6/ "Stereolithography Automates Prototyping", by Daniel Deitz
Mechanical Engineering, Februar 1990

/7/ "SLA USER REFERENCE MANUAL",
3D systems, inc.

/8/ "Test of Elevator Positioning Accuracy",
Danish Technological Institute,
INSTANTCAM, Workarea3, Workpackage 9

/9/ "Reduced Distortion In Optical Freeform Fabrication With UV Lasers",
by E.J. Murphy, R.E. Ansel and J.J. Krajewski, DeSoto, Inc., Des Plaines, Illinois,
Radiation Curing, Februar/may 1989

/10/ "Chain Tolerances and Accuracy in SLA Building Process",
Diplomarbeit von Leif Störmer am BIBA 1991

/11/ "Stereolithographie - ein direkter Weg zur Herstellung von Unikaten und Prototypen"; Prof. Dr.-Ing. Bernd E. Hirsch, Dipl.-Ing. H. Müller
Jahrbuch f. Optik u. Feinmechanik 1991; Schick u. Schön GmbH Berlin

/12/ "The present state of accuracy in Stereolithography"
Jan Richter, Dr. Paul Jacobs; Second international conference on Rapid Prototyping; June 23-26 1991; Dayton, OH

PAPERS RECEIVED AFTER DEADLINE

Knowledge based process planning for one-of-a-kind production

E. Hämmerle, H. Bochnick, B. E. Hirsch and J. Opas

Bremer Institut für Betriebstechnik und angewandte Arbeitswissenschaften, Universität Bremen, Postfach 330560, D-2800 Bremen 33, Germany

Abstract
This report briefly explains the problems in process planning for one-of-a-kind production. One possible approach towards an improvement will be described. Integrating kernel element is the feature based product and process planning representation. This approach is currently under development in a BRITE/EURAM research project called "Manufacturing Cell Operator's Expert System (MCOES)". First experiences and results will be presented.

1. INTRODUCTION

The emphasis in European industry is shifting away from the increasingly congested and competitive mass production arena toward small series, customer-oriented production, and what is termed 'one-of-a-kind' production. More and more companies will be confronted in the near future with the requirements of one-of-a-kind production. The conservation and development of skills to fulfil these requirements and the ability to manufacture one-of-a-kind products will determine the companies success. Process planning as part of operative production engineering is a key factor for that success. Therefore appropriate methods and computer supported tools have to be developed.

2. REQUIREMENTS

One-of-a-kind products are in principle products which will be produced only once (in a life time). They are mainly very complex products like ships, power plant stations, bridges, planes and special tool machines. Basically in the prototype stage any product can be called a one-of-a-kind product. So even batch manufacturers have to temporary fulfil the requirements of one-of-a-kind production. Table 1 lists some characteristics and distinctions between one-of-a-kind-, single- and batch production.

Product complexity, production complexity, production and planning flexibility are the main impacts on the decision, planning and manufacturing principles and lead to the briefly summarised general requirements:

- Customer influence during the whole product life cycle requires a high planning and production flexibility for adaption to possible changes.

Table 1.
Characteristics of one-of-a-kind-, single- and batch production.

criteria	one-of-a-kind production	single production	batch production
customer oriented production with redesign	++	−	−
simultaneous engineering	++	+	−
deficit of relevant manufacturing information at order entry	++	−	−
just in time generation of product information	++	+	−
construction site and shop floor manufacturing	++	+	−
reconfigurable production equipment	++	+	−
qualification and human interaction are main factors	++	+	+
virtual production island is an alternative organisation type	+	+	−
manufacturing with batch size one	+	++	−
manufacturing based on an already existing design	+	++	++

++ yes,   + partly,   − no
Reprinted from: CIM-TT Bremen [1]

- Enterprise activities are not exactly predefined and synchronized and executed rather in parallel than in sequence especially product design, process planning and production planning.
- Decision management based on incomplete information especially in the early product life cycle stages.
- Support of different kind of resource organisations and different levels of automation.
- Concurrent engineering support because of multi-discipline engineering and distributed manufacturing concepts.

Solutions for one-of-a-kind manufacturers have to fulfil these requirements and have to be directed towards organisational and technical improvements. Group technology aspects, the creation of decentralized autonomous organisational units together with the reintegration of planning and control tasks on shopfloor level are organisational approaches to secure and support the strengthening of human competence and experience. Technical solutions in the form of appropriate computer systems and software encourage the positive effects of organisational improvements. Technical solutions for process planning will be explained in more detail in the next chapters.

3. PROCESS PLANNING

Process planning is a subfunction of operative production engineering. Its objective is the determination of processes and parameters, the sequence of processes, the work content and the required resources necessary to convert a part from its initial form to its final one.

Appropriate system oriented solution approaches for one-of-a-kind process planning are

- an experience based system development,
- a modern object oriented software engineering environment,
- a feature based product model common to all process planning modules,
- a knowledge based representation method,
- a functional integrated system for method-, setup-, tool- and fixture planning,
- an hierarchical process plan structure to represent different process planning levels with a different degree of process planning knowledge and with the possibility to decompose activities,
- an effective process plan representation of parallel, alternative activities.

In the following chapter we will refer to a research project trying to fulfil some of these requirements by an integrated process planning system.

4. THE MCOES PROJECT

MCOES (Manufacturing Cell Operator's Expert System) is a 3-year European project under the BRITE/EURAM program of the Commission of the European Communities. It started in June 1990 and is now at the end of the prototype phase. The main aim of the project is to improve the efficiency of factories producing small batch and one-of-a-kind products. This is to be achieved by using expert system techniques. The MCOES system will produce all of the workshop information (e.g. routing, tool selection, fixture plans, cutting strategies) necessary to manufacture a part. This also includes the generation of NC code for Computer Numerical Controlled (CNC) machines present in a workshop or Flexible Manufacturing System (FMS).

By giving the part designer immediate feedback and suggestions of alternative design options, MCOES drastically reduces the time taken in consultation with the manufacturing facility. This is very important in one-of-a-kind production, where a short lead time and fast re-tooling and setup are of paramount importance to retain productivity.

Gains in productivity are achieved by integrating several processes on the path from design to manufacture into the MCOES system. Together with solving some processes in parallel the MCOES architecture is construed for one-of-a-kind process planning as shown in figure 1.

As seen in figure 1 the MCOES system consists of two parts. One is to support design level, while the other is to support process planning level. The part designer should be informed about errors in parts as well as about manufacturing friendliness and complexity of part. The manufacturing engineer should be supported in fixture planning, tool planning, method planning as well as in NC code generation. To develop an interactive expert system

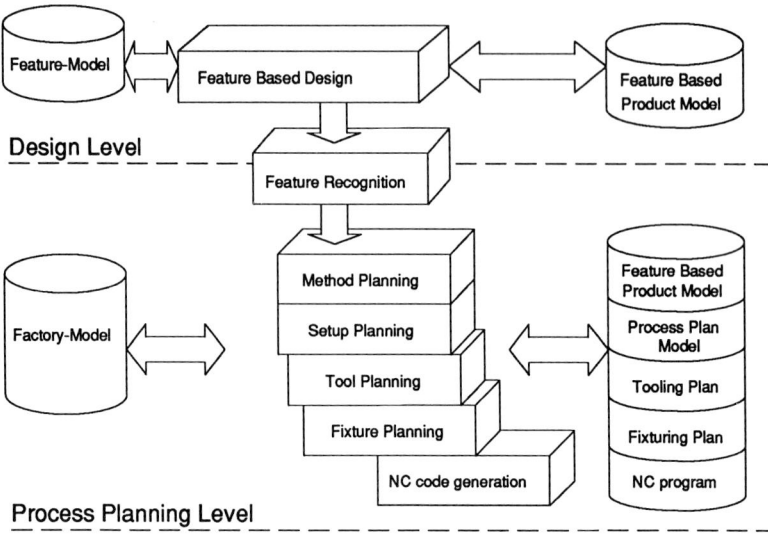

Figure 1. The MCOES system architecture.

that integrates these objectives, knowledge representation methods and data structures for a product model and process plans are needed. They are briefly explained in the next chapters.

4.1. Knowledge Representation

The know-how of a process planner can be classified into recognition type know-how and procedural type know-how. Recognition type know-how is required to recognise the part shape, status and condition. Procedural type know-how is dedicated to decide manufacturing process selection and how to react on status and condition.

In MCOES two of the 'standard' knowledge representation techniques are used: frames for representing declarative knowledge while production rules and handlers are formalising procedural knowledge.

Frames are data structures in which all the knowledge about a particular object or event is stored. Such an organisation of knowledge is useful for modularity and accessibility of the knowledge. In addition the MCOES frame system [2] allows to specify default values for attributes of an object. Each object is described by some attributes (slots), containing information to process planning activity. The objects can be hierarchically structured. So defined attributes can be inherited from superclass which has the role to define them. Frames are also useful to describe relationships between objects, like 'is-a-part-of'. It gives the possibility to define semantic networks.

Handlers are a feature of the MCOES frame system. Handlers are Lisp functions that allow message passing between frames. Handlers are assigned to frame slots as values, the name of the slots being the same as the messages. Therefore handlers can be shared by frames through slot value inheritance.

The procedural type know-how is generally represented by *rules* as follows:

if workpiece is a casting part
then use ribbed fixture elements for support.

The rules make use of the knowledge stored in frame slots to draw conclusions in the form of values for slots of further frames. The rules are grouped in contexts; i.e. subsets referred to the solution of particular aspects of the problem at hand. Examples are rule sets for tool selection and process selection. The partition of the set of rules simplifies the modification and extension of the knowledge base as well as the separation of declarative and operative knowledge.

4.2. Product model

The product model of the MCOES project is based on a *feature model representation* [3]. No product is entirely the same as a previous one. But characteristic regions of interest within a product can be modelled by features. The following features are defined: *design features, manufacturing features and fixturing features.*

Different experts have different views and information about a product. For a designer a design feature reflects a functional view, while for a process planner a manufacturing feature reflects a manufacturable view. For instance, the designer deals with holes which process planning later reasons should be drilled, bored or reamed based on associated tolerances.

Therefore it has to be distinguished between those two kinds of features. Nevertheless a translation of design features into manufacturing features is necessary to bridge the gap between those different views for example by rule based methods. All kind of features are represented by frames in MCOES.

The task of manufacturing features is to provide information needed for automatic process planning (selection of manufacturing methods, machine tools, cutting tools, etc.). In addition manufacturing features have to include geometric information needed for NC code generation.

Features added to a workpiece to allow easier fixturing are called fixturing features. For automatic fixture planning some features of a product model have to be identified as fixturing features.

Features can be grouped into classes and arranged into a class hierarchy. MCOES distinguishes between basic-, container-, surface-, billet-, texture and tolerance features. Basic features for example are then further divided into holes, rotational- and prismatic features. The leaves of the hierarchy are so-called types. The other nodes are classes. Each type and class is described by a frame. Types are the only nodes which can have instances. An instance inherits information from its type and higher level class frames. In the hierarchy different types of features can be found. The information related with the features of a product are essential for automatic process planning.

For features there is also a need to distinguish between different levels of information. *Generic features* are a general parametric description of geometry and manufacturing attributes common to a certain class of features. Values of slots are not known. Generic features are company and order independent.

Variant features are a subset of generic features. Values of slots are specified. Parameter values are either restricted by national/international standardization activities (DIN, ANSI) or company specific standards. Variant features can therefore be further classified into company dependent and company independent variant features.

One could say, variant features are 'manufacturable' features. They are one possible way to represent manufacturing knowledge, especially company specific knowledge, in combination with variant methods and variant work elements.

A work element (we_1 and we_2 in figure 2) is defined as an execution of operation for one feature by one tool. Work elements are usually executed at machine or equipment levels of factory control hierarchy.

The idea of a variant feature is strongly related to the variant planning concept. If one knows a certain dimension of a diameter the respective manufacturing methods and tools might already be known. In that case variant features and variant work elements are predefined, reusable templates which simplifies process planning.

4.3. Process plan structure

There are a number of issues common to all process planning applications. These include the representation of process plans and the definition of an architecture which can communicate and control these representations internally and with other systems of production engineering. The definition of these issues is called the definition of the process planner structure [4]. A standardised process plan representation called ALPS was chosen as the base of the MCOES project [5]. Functionally the structure is very similar to an ISO model for the process planner that is currently under development.

A process plan can be thought of a collection of individual processes organised into a coherent structure directed toward accomplishing some goal. The structure for relating these processes in accordance to the above mentioned one-of-a-kind production requirements must allow for:

Process precedence:
> determine the sequence of task.

Alternative sequences:
> express different task sequences which provide a means for the scheduler to determine which sequence is currently optimal.

Parallel actions:
> explicitly show how multiple task sequences within a plan can be performed at the same time. It is assumed that separate plans can be executed in parallel as separate jobs.

Decomposition of activities:
> support the concept of hierarchical plans, where a process task at a higher level can expanded into a collection of processes at a lower level.

Synchronisation:
> provide for synchronisation between multiple parallel task sequences within a plan (as in item 3) and between plans.

Resource monitoring:
> provide the means for collecting and updating statistics for resource availability and utilisation to support scheduling and resource allocation.

Extensibility:
>support extensibility by not constraining the user fixed functionality.
>Users must be able to customise process plans to support this facility.

Instead of using the original graph based representation of ALPS, MCOES has chosen hierarchical lists (see figure 2).

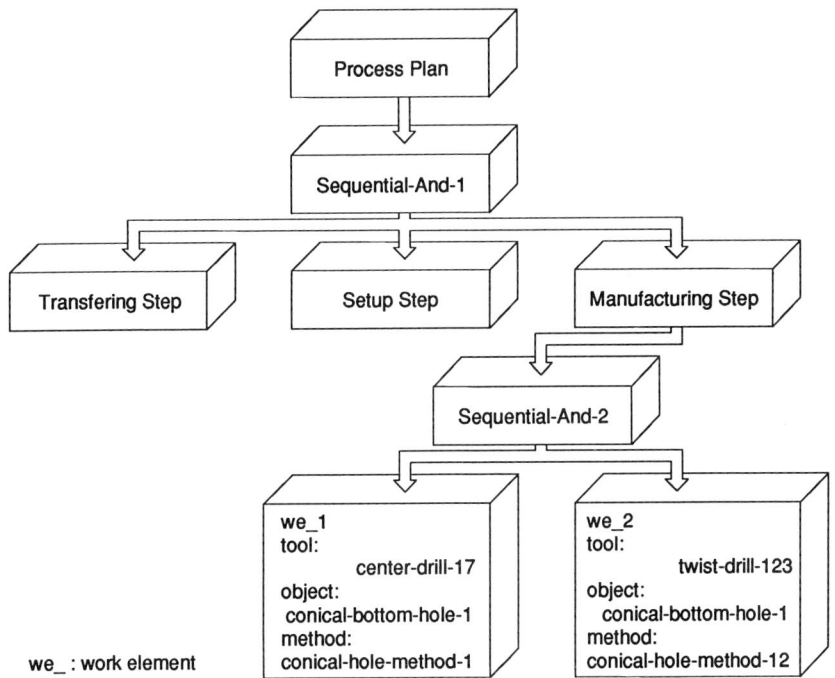

Figure 2. Sample process plan.

In the prototype phase MCOES will concentrate on variant planning; hence parts should consist of variant features only. Variant features include previously coded process planning knowledge. This information is included in the respective class frames by using the plan specification syntax. Plan specifications describe the planning logic and refer to the required steps and methods. When process planning is performed, the plans expressed through the specification syntax are elaborated.

4.4. Knowledge Engineering Environment

Primary programming languages of the MCOES project are Common Lisp and C. In knowledge engineering applications Lisp is the most common programming language. The advantages of Lisp are availability of object-oriented extensions, comfortable environment allowing easy software development and debugging. C is used for those software modules for which the use of Lisp is not so suitable for technical or other reasons, for example geometric modelling.

The data Representation language in MCOES is the BEEF frame system. It represents the objects identified in process planning using the concept of frames. Graphs of frames can be formed by using links, which are ordinary formed using slots with reference to a frame. BEEF provides handlers which are Lisp functions attached to classes, too. The reason for using BEEF instead of some other object-oriented representation language are the possibility to define non-standard inheritance behaviours and availability of a frame editor.

The adaption of an object-oriented database management system in the second phase of the project is on discussion because BEEF lacks support for persistent storage. BEEF frames are currently stored in text files.

MCOES contains a rule language using BEEF frames as its data elements. A data element is matched to the condition parts of a set of rules. The result of the action parts of the matching rules are returned as a list in the order of the rules.

Acknowledgement

The work described in this paper has partly been founded by the Commission of the European Communities in the frame of the BRITE/EURAM project 3528 MCOES. The authors would like to thank all MCOES partners for their contribution results partly presented in this article.

5. REFERENCES

[1]: Prof. B. E. Hirsch, CIM in der Unikatfertigung und -montage, Band 16, TÜV Rheinland, Köln 1992 (to be published 1992, internal pre-version).

[2]: Ora Lassila, BEEF Reference Manual, Helsinki University of Technology, Faculty of Information Technology Department of Computer Science 1990.

[3]: Specification of Feature Model, BRITE/EURAM, Project no 3528, MCOES WP4-D3(R), May 1991.

[4]: Process Planer Structure, BRITE/EURAM, Project no 3528, MCOES WP6-D1(R), March 1991.

[5]: S. R. Ray, A Modular Process Planning System Architecture, IIE Integrated System Conference, Atlanta, Georgia, November 1989.

PROBLEMS IN DESIGNING HIGHSPEED MILLING TOOLS

Prof. Dr. -Ing. H. Schulz
Dipl. -Ing. W. Hahner
Dipl. -Ing. U. Rondé

Institute for Production Engineering and Machine Tools (PTW)
Technichal University Darmstadt

ABSTRACT

High Speed Milling needs specific machine components, especially concerning the tools and their interfaces to the motor-spindle.
The developed systems have been successfully tested during several joint projects. The PTW is carrying out standardization work on High Speed Milling Tools focusing on product safety considerations. In order to optimize the tool design, the Institute for Production Engineering and Machine Tools has developed a modified Failure Mode and Effects Analysis concept. A control loop model describes the data flow in the design phase of a product and the relevant points of decision. Combined with a documentation strategy an integrated safety concept is achieved.

Key words: High speed Milling, Milling Tools, FMEA, Design Rules, Destructive Testing, Standardization.

1. INTRODUCTION

The reduction of the essential processing time is an important objective for cutting intensive machining operations which can be achieved by raising the cutting speeds. Modern cutting materials and progresses in developing machine tools and their components allow a considerable reduction of the essential processing times. Further advantages of the High Speed Cutting Technology are high surface qualities and low cutting resp. passive forces.
During the last years the Institute for Production Engineering and Machine Tools developed new concepts for machine tool components by intensive researches. High speed main spindles, highly dynamic drive- and control systems, fast moving machine parts with small masses, clamping means, security devices as well as new tools were designed, built and tested.

2. DESIGN OF HSC TOOLS

The selection of the cutting material and the optimization of the cutters geometry and the entire design of the tool has to be adapted to the conditions of high speed cutting. Besides the cutting forces the tool is stressed especially by high centrifugal forces during the milling operation. The cutting forces produce a surface load which is built up on the contact zone of the edge. At appropriate cutting geometry the cutting force presses the indexable insterts into their seats so that the clamping elements as screws and daws are not additionally stressed.

Concerning the clamping elements and the tool body, basic safety criteria have to be considered because of the high centrifugal forces.
The occuring centrifugal forces are able to destroy the tool body and the fitting of the indexable inserts. Stress calculations on disc like rotors of different materials have been carried out. They showed that, compared to steel tools, one can achieve higher speeds with materials of lower specific weight.
A condensed list of requirements for the High Speed Milling Tools can be described as follows:

a) high mechanic stiffness.
b) rotationally symmetric structure in order to avoid balance errors.
c) minimum radial deviation under centrifugal forces
d) elements to adjust balance errors

Conventional stress calculations on ring geometries give a good view concerning the effect of centrifugal forces on fast rotating tools. The diameter of the center whole strongly influences the limit of rotational speed (Fig. 1).

Because the cutting speeds depend on the workpiece material quite different tool concepts exist. Tools which are close to conventional series can be used for the high speed machining of steel and cast iron materials. For the machining of aluminium, copper alloys or fibre reinforced plastics the PTW developed special tools in cooperation with industry. Overspeed tests have shown, that conventional clamping systems are suitable for cuttig speeds up to 2000 m/min (Fig. 2). At higher cutting speeds, interlocking connections between inserts and main body are necessary. Fig. 3 shows some developed clamping systems.

3. INTERFACES BETWEEN TOOL AND SPINDLE

The result of high speed machining essentially depends on the spindle-tool interface and on the corresponding clamping system, because these components have to fulfill their functions under extreme conditions. An optimized design of this interface is a basic need not only for the fast, automatic exchange and for high transferable powers but also for highest accuracy. The general requirements for these interfaces during a milling operation are as follows:

- Reliable transfer of forces and moments.
- Suitability for manual, semi-automatic and automatic tool exchange.
- High repeating accuracy in tool exchange.
- Maximum static and dynamic stiffness.

Additional requirements for the High Speed Cutting Technology are:

- Little radial deviation at high speeds.
- Increased concentricity and run out accuracy.
- High positional accuracy.
- Short tool changing cycles.
- Reduced influence of centrifugal forces by small diameters and low masses.

For High Speed Milling two different interfaces succeeded: the Quick Release Taper and the draw in collet chuck.

The Quick Release Taper chuck fulfilles the essential demands and is therefore used in many applications.
Regarding the HSM-Spindles the tools are equipped with a steep angle taper (Quick Release Tapers), DIN 69871, and are connected with the spindle by different clamping systems. Because of the high occuring centrifugal forces there is an elastic deformation of fixture and tools (Fig. 4). Here the expansion of the fixture is bigger, causing an axial tool movement towards the spindle. The resulting clamping forces after rotation are very high. An automatic tool change nearly gets impossible. The elastic deformation is strongly depending on the spindle design. The non parallel deformations cause low accuracy of the tool positon, especially when dynamic milling forces are present. The resulting excentricity and the axial displacement lower the obtainable surface qualities. Depending on the different process conditions, the clamping forces may vary. In the worst case the rotation tool could slip out of

the spindle, causing serious damage. During a joint project in cooperation with industry on High Speed Milling the tool-spindle interface problems have been investigated. Various design proposals have been tested in order to optimize the tool clamping. An example of a design modification is shown in figure 5. The circular plane surface limits the axial displacement of the tool to the remaining airgap between tool and spindle. The forces that are needed for changing the tool can be limited in this way. The disadvantage of this solution may be localized in the axial positoning of the tool by two functional elements. Besides this a polygone equipped steep angle taper has been developed, where comparatively high torque can be transferred (Fig. 6). Figure 7 shows aquite new tool-spindle interface which is tested in the PTW laboratories. The conventional clamping of cylindrical tools with collet chuck is not preferable when working at high rotational speeds. The centrifugal forces reduce the clamping forces. The clamping nut can even start turning by itself during a spindle deceleration. Special high speed clamps have been developed during the project on High Speed Milling. Their significant properties are high circular accuracy and extended axial clamp length with high clamping forces. The clamping nut is equipped with ball bearings and is integrated in the collet holder.

4. DESIGN AND SAFETY OF HIGH FREQUENCY TOOLS

The High Speed Milling Tools must be designed to guarantee the safety of the personnel and potential damage on machine tool components in the vicinity of the rotating tools must be minimized. The list of requirements for High Speed Milling machines concerning safety standards is more sophisticated than in conventional milling machines. The energy level in case of failure is much higher. The sudden failure does not allow the operator to interfere. Therefore the PTW work is focused on active-safety -design, besides passive, computer-aided surveillance devices. For the prediction of potential failure the FMEA (Failure Mode and Effects Analysis) is applied successfully. The FMEA method itsself has been optimized in order to meet the specific needs of the machine tool industry. Potential failure can already be estimated during the design phase. The machine is structured according to systems and subsystems, single parts are analysed even down to the surface level. The FMEA selcts the components which are to be tested under real-life conditions. This strategy allows minimum cost testing.

5. MODIFIED FMEA FOR SAFE HIGH SPEED TOOLS

Product design is a complex form of data processing. Every design step requires multiple decisions, according to the informations available. All together they determine the final quality of a product in terms of reliability and safety (Fig. 8, 9). It is necessary to know the data flow, in order to determine critical design steps. The PTW has developed a control loop model of the design process, which describes the flow of relevant data. It includes a model of the product structure, derived from existing theories. This model is used to analize the design quality of different High Speed Milling tools and leads to individual testing procedures.
Each step of development needs specific concepts to verify that the overall requirements are met. The control loop model gives a reference to design the different kinds of analysis (Fig 10).
The documentation of results is carried out according to the structure given by the model.
A homogeneous documentation system is essential to any step of design review especially referring to legal requirements on safety relevant products. Using the control loop model, the FMEA has been modified in several points. Its application thereby gets more efficient and more precise. The control loop model is very useful for the development of software modules, which can help the design engineer to optimize his work. It is fundamental for an Integrated Quality Assurance Concept in product development and for any kind of design reviews.

6. STANDARDIZATION WORK

The PTW has been charged by industrial partners to develop safety regulations for high speed milling tools (fig. 11). It is running a test bench which is able to accelerate tools up to 200.000 1/min in order to determine the maximum rotation speed and the process of destruction induced by centrifugal forces. A tool diameter of 200 mm and a tool length of 240 mm can be tested up to a weight of 10 kg (Fig. 8). The target of this project in cooperation with actually 10 industrial partners, is the constitution of international design and test standards for High Speed Milling tools. Especially the questions concerning insert clamping mechanisms and centrifugal forces in tool bodies are analysed in order to give recommendations for design.

REFERENCES

Schulz, H.; Hochgeschwindigkeitsfräsen metallischer und nichtmetallischer Werkstoffe, Hanser 1989

Müller,M.; Hochgeschwindigkeitsfräswerkzeuge. Tagungshandbuch, 4. Darmstädter Fertigungstechnisches Symposium, 1989 TH Darmstadt.

Walz, T.; Schneider, M.; Hahner, W.; Mootz, A.; Aktive und passive Schutzmaßnahmen beim HSC-Fräsen. Tagungshandbuch, 4. Darmstädter Fertigungstechnisches Symposium, 1989 TH Darmstadt.

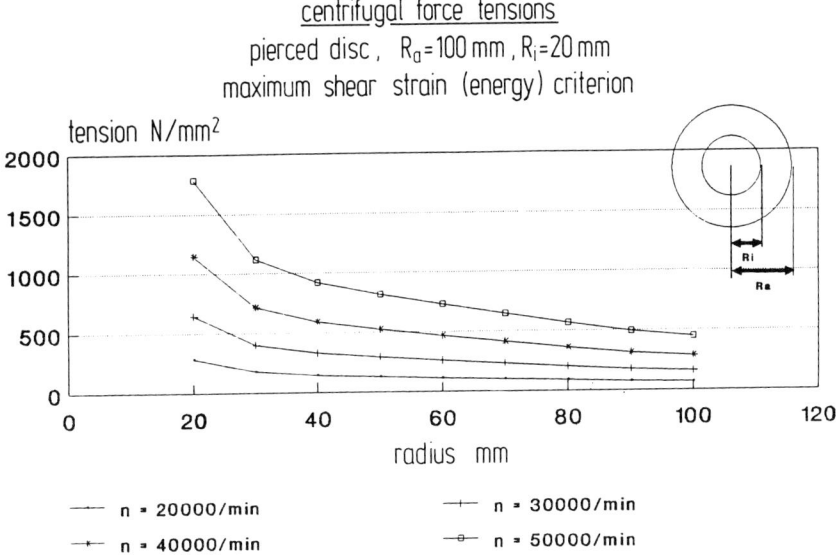

Fig. 1. HSC-tool-safety

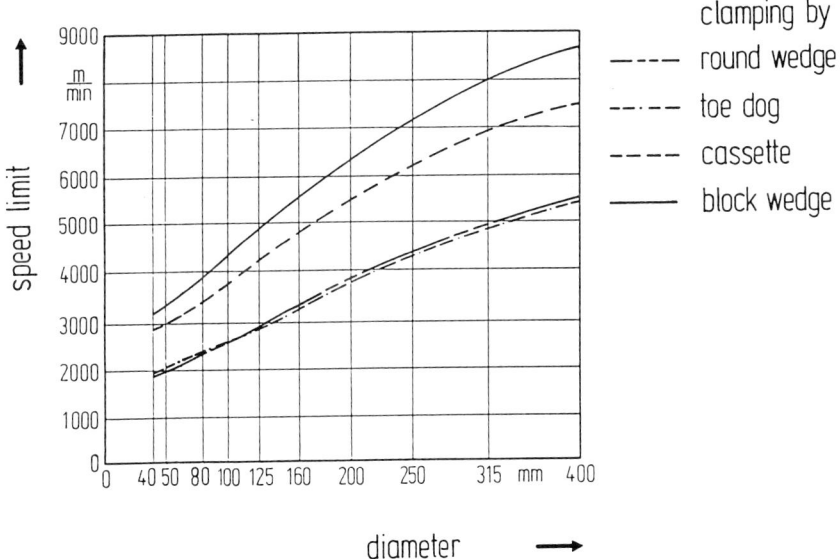

Fig. 2. Limits of cutting speed for different insert clamping systems

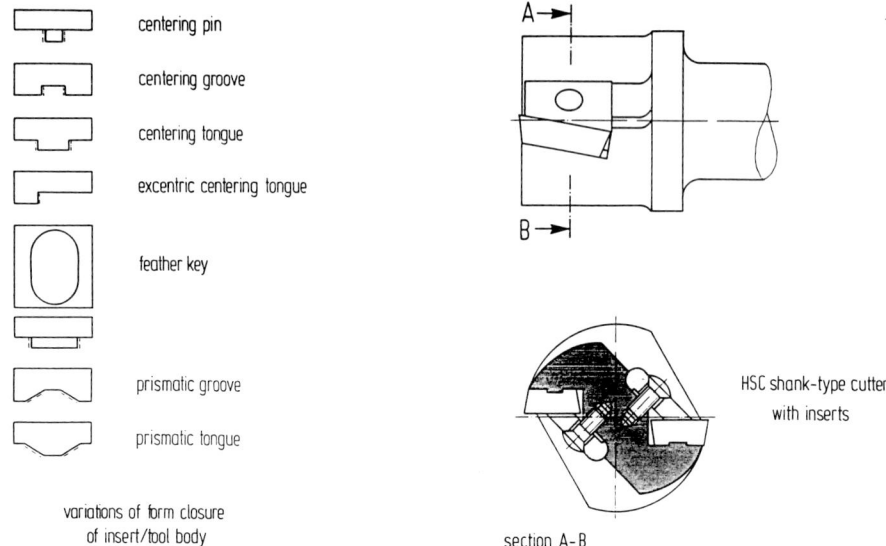

Fig. 3. High speed tools insert clamping

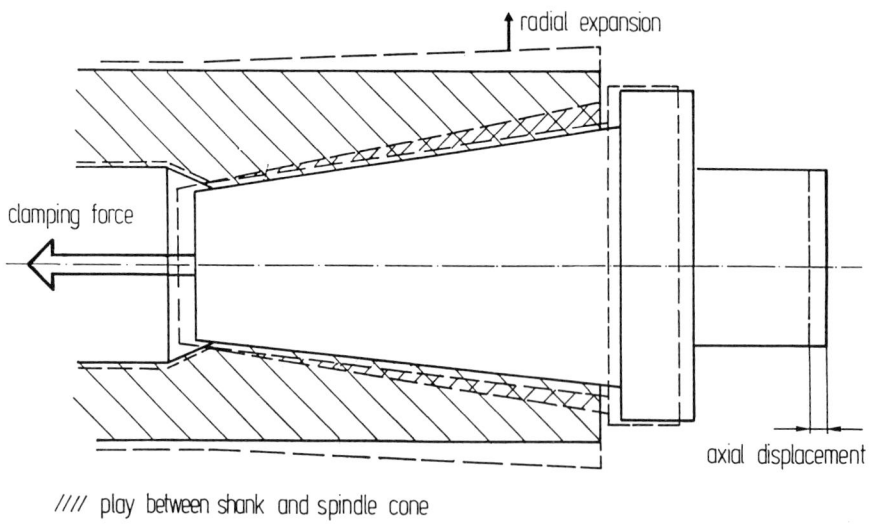

Fig. 4. Expansion of spindle cone at high speed

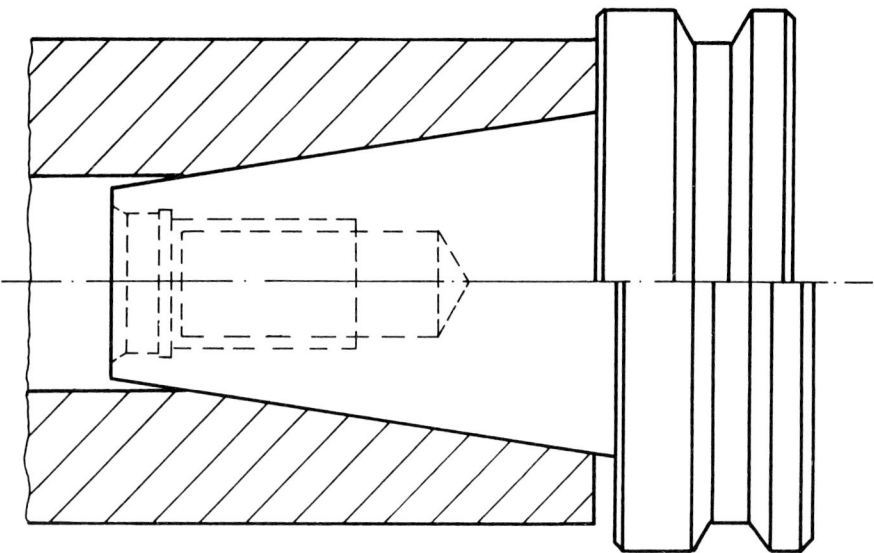

Fig. 5. Interface steep angle taper with or without facing stop

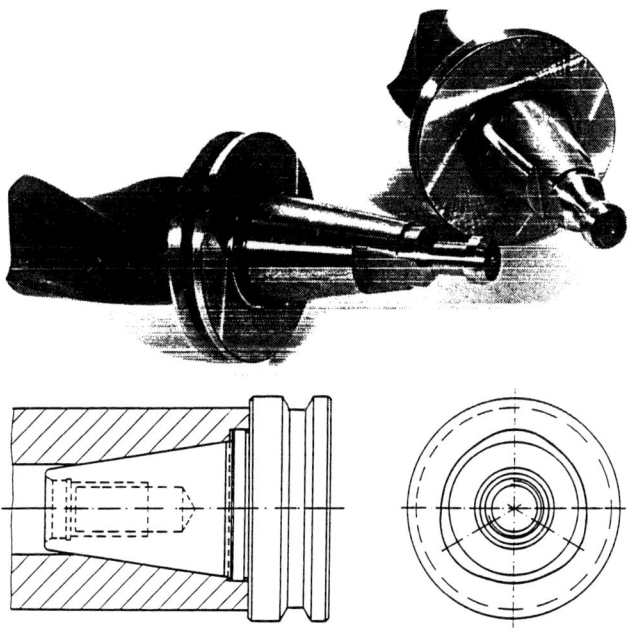

Fig. 6. Polygon equiped steep angle tape

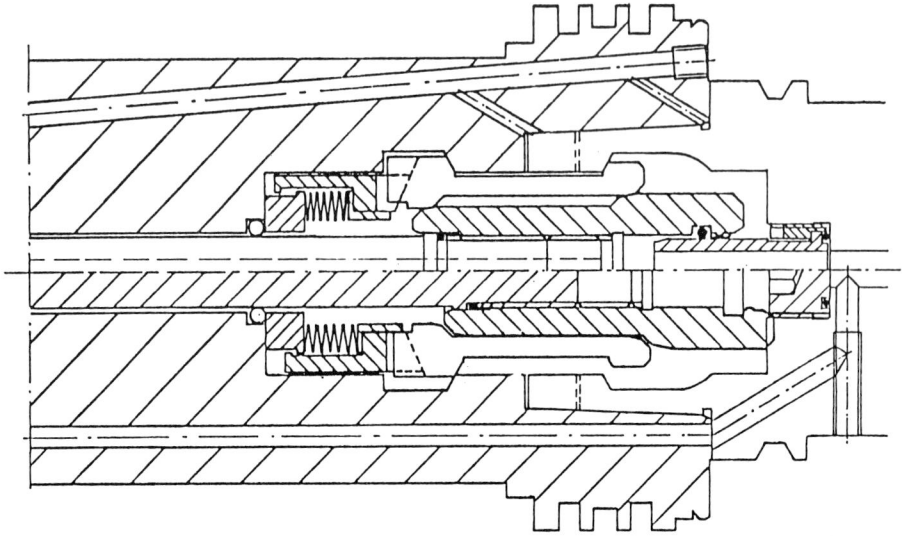

Fig. 7. New tool spindle interface

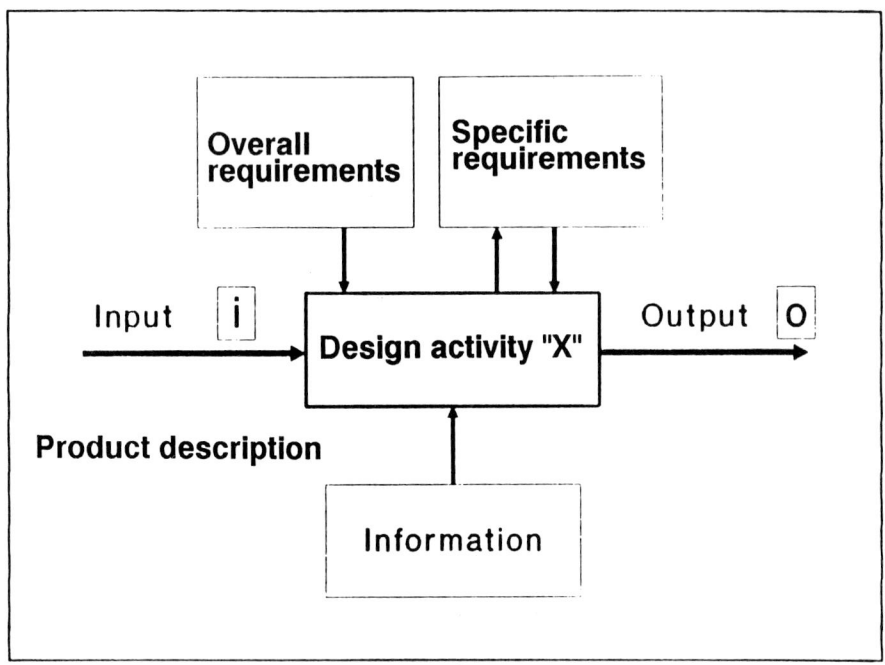

Fig. 8. Information in design

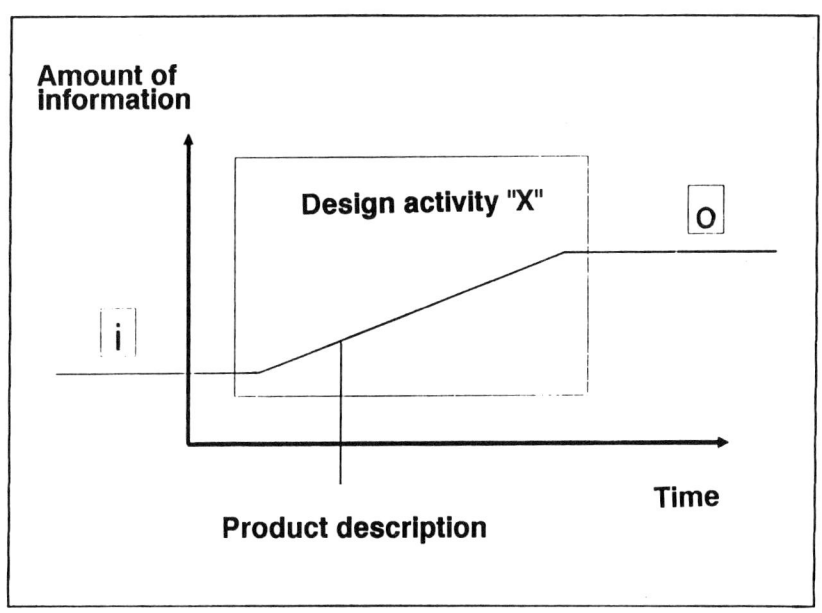

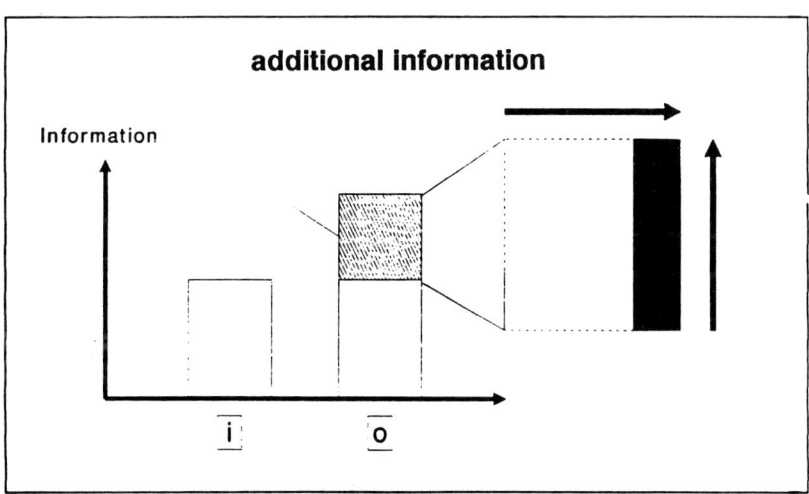

Fig. 9. Design steps and information

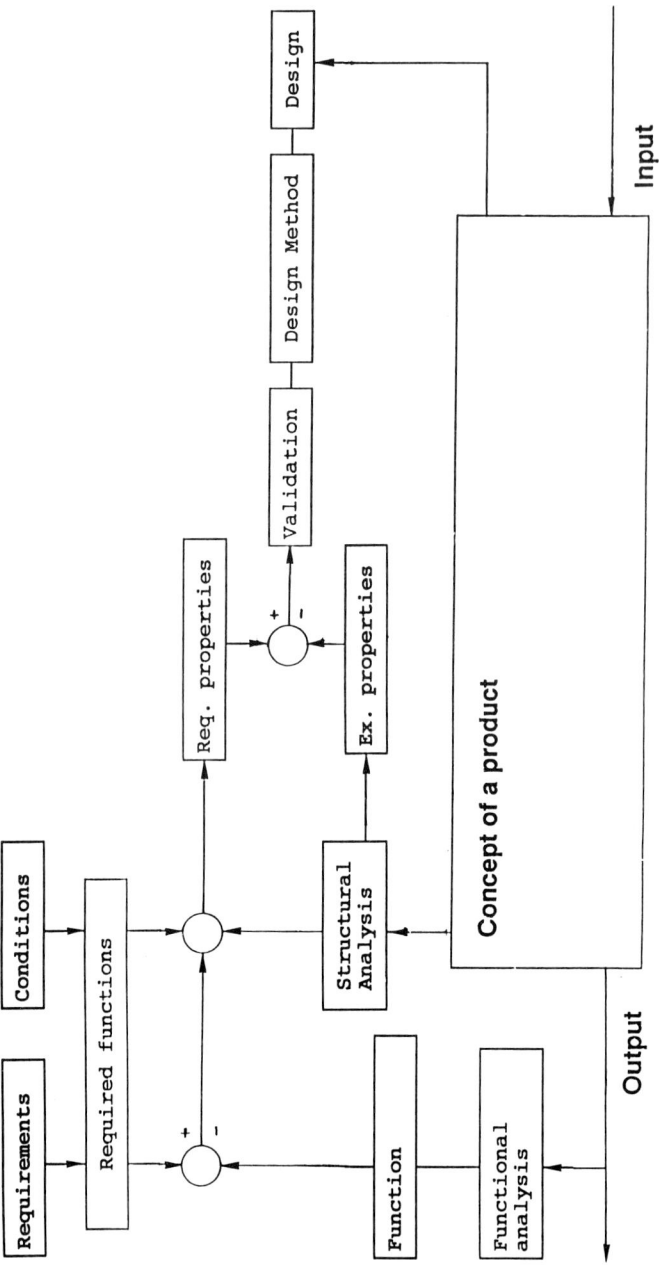

Fig. 10. Control loop model

max. sample diameter d = 200 mm
max. sample length   l = 250 mm

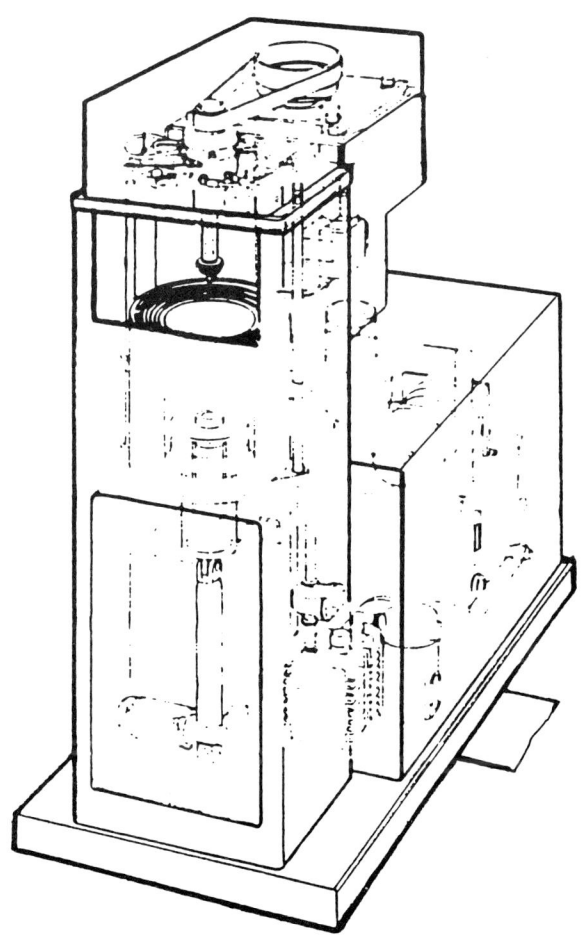

Fig. 11. Overspeed test system for high speed rotors